MILKY WAY

From 13.5 billion years ago to the infinite future

© Haynes Publishing 2019

Gemma Lavender has asserted her right to be identified as the author of this work.

First published in November 2019

A catalogue record for this book is available from the British Library.

ISBN 978 1 78521 590 2

Library of Congress control no. 2019942472

Published by Haynes Publishing,
Sparkford, Yeovil,
Somerset BA22 7JJ, UK.
Tel: 01963 440635
Int. tel: +44 1963 440635
Website: www.haynes.com

Haynes North America Inc.,
859 Lawrence Drive, Newbury Park,
California 91320, USA.

Printed in Malaysia.

MILKY WAY

From 13.5 billion years ago to the infinite future

Owners' Workshop Manual

An insight into the study of our home galaxy
and our place in it

Gemma Lavender

Contents

OPPOSITE A star throws off its layers of gas and dust, leaving behind a planetary nebula – a common resident in the Milky Way as its stars evolve. Pictured is Abell 24 in the constellation of Canis Major (the Lesser Dog). *(ESO)*

Introduction

If you are lucky to live under a reasonably dark sky, or get the chance to visit such a location, then you will see the beautiful shimmering spectre of the Milky Way arcing overhead. It looks like a river of light, but train a telescope on part of it and you will see that it is constructed from countless stars. Look even closer and you will see that among the stars are pockets of gas and dust, some colourful, others dark and hidden.

This river of stars, gas and dust is our Milky Way Galaxy. It doesn't look like a galaxy to us, but that's because we're inside it, and we're looking through it, edge-on. In reality, our Galaxy is a vast disc, approximately 100,000 light years across, and our Sun is but one tiny speck, one of 200 billion in the Milky Way.

To give an idea of the scales involved, consider that the closest star to us, which is Proxima Centauri, is 4.2 light years away (39.7 trillion km/24.7 trillion miles). If we could scale down the Galaxy and everything in it, so that our Sun, which is 1.4 million km (863,700 miles) in diameter, was the size of a grape, then Proxima Centauri would be 560km (350 miles) away, and the entire Milky Way Galaxy would stretch 12.9 million km (8 million miles) across.

The vastness of our city of stars is placed into context when we consider the larger universe. The Milky Way is just one of an estimated 150 billion galaxies in the observable universe, which is a volume 93 billion light years across. It many ways, it's a fairly typical spiral galaxy, but it also has some unusual characteristics that set it apart – for example, the black hole at its centre is quite small, comparatively speaking, compared to other galaxies, which hints at our Galaxy's mysterious

history. And by understanding the Milky Way better, we can come to grasp in more detail the properties of other galaxies, how those other galaxies formed, and how the Milky Way compares with other galaxies in this cosmic collection of island universes, which in turn helps us to better understand the Milky Way, and so on.

Equally important is the fact that the Milky Way is our home, one that we can see whenever we go out at night and look up at a cloudless sky. Every single star that we can see with our eyes when we look up also calls the Milky Way home, and the Sun is their neighbour. There's so much to see and explore in the Milky Way, but in this book we'll give you a taste of what our galaxy has to offer.

In the first chapter, we'll look back at some of the major milestones in history that played key roles in developing our appreciation of the fact that we do live in a spiral-shaped galaxy, one of many in the universe. From Galileo Galilei to William Herschel, and from Johannes Kepler to Edwin Hubble, we'll meet some of the big names in science that made these historic astronomical contributions.

In the second chapter, we'll go back to the beginning of everything – the Big Bang – and chart the history of the universe from that moment, and how everything fell into place for the Milky Way Galaxy to be born and to subsequently develop its spiral structure.

Then in chapter 3, we'll meet the Milky Way's family, called the Local Group of galaxies, which includes the famous Andromeda Galaxy, and the nearby Large and Small Magellanic Clouds, and discover what similarities and dissimilarities they have with the Milky Way.

Following that, we'll dive headlong in to the Milky Way in chapter 4, starting with the Sun's immediate neighbourhood, and the stars and clouds of gas that reside within it. Then, in the next chapter, we'll explore the telescopes on the ground and in space, collectively covering the entirety of the electromagnetic spectrum, that we use to explore the wider Milky Way. In chapter 6 we introduce some of the most fantastic inhabitants of the Milky Way and explain what they are, from the awesome Orion Nebula to the most ginormous globular cluster.

Finally, in chapter 7, we will reveal the fate of the Milky Way, as it eventually crashes into the neighbouring Andromeda Galaxy and begins the final stage of its evolution.

BELOW A fish-eye view of our galaxy. Four 8.2m (26ft 11in) Unit Telescopes can be seen, along with four 1.8m (5ft 11in) diameter Auxiliary Telescopes, making up the Very Large Telescope. *(ESO/Y. Beletsky)*

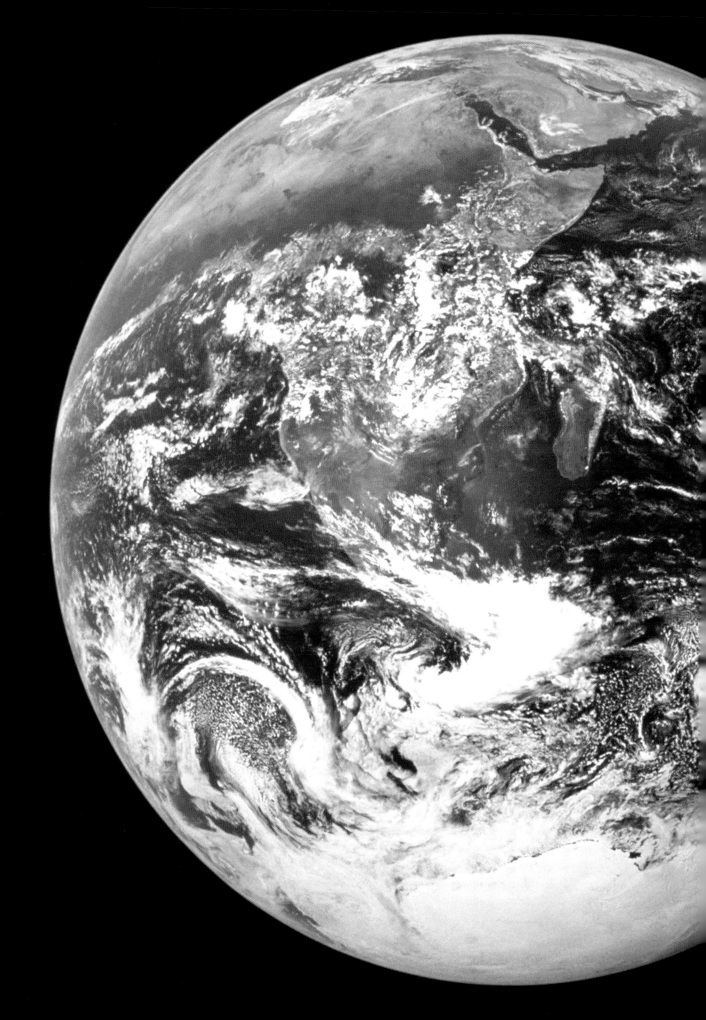

Chapter One

Our home galaxy

Earth: our home among a family of gaseous and terrestrial planets, as well as lumps of space rock of varying shapes and sizes, orbiting a star. The star – of course – is our Sun; a yellow dwarf that sizzles at a temperature of over 5,500°C (9,932°F), shedding warmth and light and providing the perfect conditions to keep the ideal balance for life in our little corner of an incredibly vast universe.

OPPOSITE The famous 'Blue Marble' photograph image, which was shot on 7 December 1972 by the Apollo 17 crew, is proof that delicate worlds can exist in the right conditions in our galaxy. *(NASA/Apollo 17)*

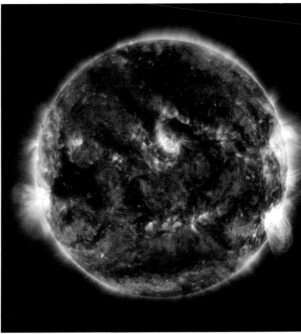

ABOVE A swarm of young and old stars in our galaxy. Such a diversity tells us much about the Milky Way's formation. *(NASA, ESA and T. Brown (STScI))*

ABOVE RIGHT Our Sun is just one of billions of stars that make up the Milky Way. Our nearest star is a yellow dwarf. *(NASA/SDO)*

BELOW An artist's impression showing the diversity of planets in our solar system. *(NASA/Jenny Mottar)*

But our nearest star isn't alone. Beyond the 1.39 million-kilometre- (860,000-mile-) diameter ball of churning plasma that is our Sun, weighing in at 2 million trillion trillion kilograms, there are other suns. Some are larger, some are more massive, some are smaller, whilst others are less hefty. Together they form a network, tightly bound by the force of gravity, and, if you move out of our solar system far enough – say, over 25,000

light years out (where one light year is 9.46 trillion kilometres or 5.88 trillion miles) – you won't just whizz past hundreds of thousands of stars, you'll see a pattern start to emerge. That pattern is the spiral structure of our Milky Way galaxy.

Our home galaxy is packed with around 200 billion stars, collaboratively tipping the scales at an impressive 1.5 trillion times the mass of our Sun and extending for up to 100,000 light years in diameter with a thickness of some 1,000 light years. Our first impression is that it's home sweet home, but what would immediately become apparent could you zoom outwards is the Milky Way's structure: a mixed-up pattern of haze and points of light warped into a winding spiral, with a blazing bright centre, stretched – or elongated – into the shape of a bar.

We know the Milky Way to be a barred spiral galaxy, but from our location on Earth, you'd be hard pressed to make that connection. It's only with countless pieces of data that we've been able to build up this picture over several decades. Head out for an evening under the stars, particularly when we leave spring behind and head into the summer, and you'll see an entirely different picture: a hazy band of light that – from our perspective – weaves through the major constellations of Orion (the Hunter),

Auriga (the Charioteer), Perseus (the Hero), Cassiopeia (the Seated Queen) and Cygnus (the Swan) in the Northern Hemisphere, and Sagittarius (the Archer), Scorpius (the Scorpion), Norma (the Carpenter's Square), Circinus (the Compass), Crux (the Southern Cross) and Carina (the Keel) as seen by observers in the Southern Hemisphere. It's a wondrous sight to behold but without the help of an optical aid, the human eye can only see

ABOVE The dusty lane of our galaxy, as visible from Earth. This section of the Milky Way is the region of sky that spans from the constellation of Sagittarius (the Archer) to Scorpius (the Scorpion). *(ESO/S. Guisard)*

LEFT One of the Auxiliary Telescopes that comprise the European Southern Observatory (ESO)'s Very Large Telescope sits under the splendour of the Milky Way. *(Y. Beletsky/ESO)*

a fraction of the billions of stars it contains; a mere several thousand.

That's the situation some of the very earliest astronomers found themselves in. Notably Italian polymath Galileo Galilei, who turned a telescope of his own making towards what he perceived to be 'congeries of innumerable stars grouped together in clusters too small and distant to be resolved into individual stars by the naked eye'. That evening in 1610 saw Galileo take to his journal, which later became published as *Sidereus nuncius*, or 'Starry Messenger'. For the very first time, the milky

strip that had been puzzling Earth dwellers for thousands of years, and which the Ancient Greeks had named *via lactea*, literally meaning 'milky way', was slowly assuming an identity. 'No matter which part of it one targets with the telescope, one finds a huge number of stars, several of which are quite large and very striking; yet, the number of small stars is absolutely unfathomable', Galileo wrote when describing what he could see of our majestic galaxy during his third observation.

There was still much to unravel about the Milky Way but it wouldn't become the focus of study again until 150 years later. The next to pick up the gauntlet was Thomas Wright, who put forward the idea that the Milky Way's stars arranged themselves into a flat region, extending across the entire sky and as far as the eye could see. To Wright, our galaxy appeared to be nothing more than some kind of an optical illusion; we were immersed in a flat layer of countless stars.

Of course, today, we know that not to be the entire truth. Philosopher Immanuel Kant (1724–1804), from his home country of Germany, had been paying attention; he believed that Wright's view of the Milky Way was incomplete and he knew exactly how to slot the last piece of the jigsaw puzzle into place. He suspected – correctly – that our galaxy is a huge, rotating body, rigid and packed with a huge collection of stars, stitched together by gravity. He likened the set-up to that of our solar system, wherein the planets are held in thrall to our Sun's gravity. Yet the way he'd imagined the behaviour of the visible stars of our galaxy was on a much larger scale.

Speaking of larger scales, Kant was ambitious and looked past the structure of the Milky Way. His tours of the night sky saw him chance across several fuzzy patches; the nebulae and far-flung galaxies that are familiar to us today but were known to him back in 1755 as 'Insel', German for Island. He reasoned that as they were too big to be a part of the Milky Way itself, they must be extremely far away. He believed that not only was our galaxy its very own universe, but these distant, hazy blobs of light were also their own universes, too. In his mind, the islands were likely to be akin to the continents found

ABOVE German-born astronomer William Herschel (pictured) and his sister Caroline Herschel believed that they could see the outer boundaries of the Milky Way in all directions. *(Lemuel Francis Abbott/National Portrait Gallery)*

RIGHT A painting of William and Caroline Herschel shows William polishing a telescope mirror. *(A. Diethe/Wellcome Trust)*

BELOW The very first sketch of the Milky Way – the result of a star count by astronomer William Herschel in 1785. *(William Herschel/Royal Society)*

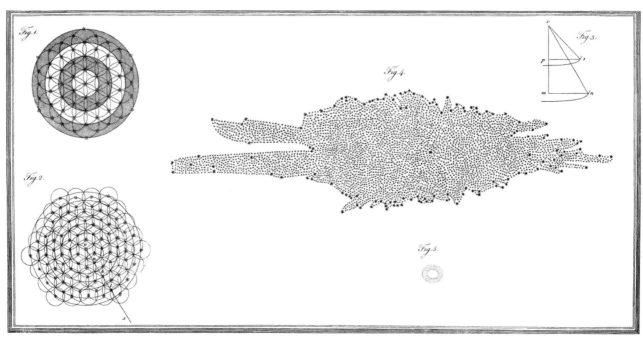

on Earth or, perhaps even more excitingly, uninhabited worlds teeming with life.

Proposals came thick and fast in the years that followed. It was in 1785 in Bath, England, at the home of the German-born but British-naturalised astronomer William Herschel (1738–1822), that the very first map of the Milky Way was painstakingly drawn by hand. It was the result of the efforts of a star count, where Herschel had sketched out the distant points of light studding the visible regions of sky. Based on previous speculations, Herschel placed our solar system at its centre, suggesting our planetary system lay close to its very heart.

The great debate

But what, if anything, existed beyond our Milky Way continued to cause a divide in opinions – especially between the greatest minds of the time, coming to a head in the form of a public debate that raged between two astronomers in 1920: Harlow Shapley (1885–1972) and Heber Curtis (1872–1942). They gathered for a showdown, albeit a calm and polite affair, at the Smithsonian Museum of Natural History, where they debated how observations of the galaxy in which we live either supported the notion that the Milky Way was the entire universe, or added fuel to the fire that the 'spiral nebulae' – Kant's 'Insel', were so-called island universes beyond the confines of our Milky Way.

The pair of astronomers were armed with scientific papers, both with opposing views. On one side of the argument was Shapley. He believed that there was nothing beyond the Milky Way; it was our entire universe and the spiral-shaped nebulae that Kant and Anglo-Irish astronomer William Parsons, the third Earl of Rosse (1800–1867), had observed through their telescopes were part of it, small objects filled with gas, studded into the star-scape. He suggested that the Andromeda spiral nebula, for example, would have to be at least 100 million light years away in order for it to exist beyond the confines of the Milky Way. Such scales throughout the universe didn't seem possible, though Shapley realised that the Milky Way at least had to be larger than previously thought to encompass all these spiral

nebulae. He used the distribution of globular clusters, which are dense spheres of hundreds of thousands of stars that surround the Milky Way, to calculate the size of the Milky Way galaxy as 300,000 light years in diameter – an overestimate of three times – and the Sun's location within it, which he calculated as 50,000 light years from the centre of the Milky Way, an overestimate of two.

Shapley also thought that he had another ace up his sleeve in the form of the highly respected Adriaan van Maanen (1884–1946), a Dutch astronomer working at Mount Wilson Observatory in the United States. Van Maanen had been watching what we now know to be the Pinwheel Galaxy (also known as Messier 101) in the constellation of Ursa Major (the Great Bear) and noticed something unusual in his results: he claimed that he had seen it rotating. If this galaxy wasn't part of the Milky Way, but was instead really millions of light years away, we should not be able to see it rotating, at least not in the way van Maanen claimed. Also, if it were rotating in such fashion, then it would surely surpass the speed of light, something impossible according to the laws of physics. The only alternative that van Maanen and Shapley could see was that, in order for us to see it rotating, Messier 101 would have to be much closer to us, a lot smaller, and inside our Milky Way.

Van Maanen also pointed out that he had witnessed a nova in the Andromeda 'nebula'. Novae – not to be mistaken for supernovae – are dramatic events whereby a small stellar core known as a white dwarf gathers so much hydrogen gas from a companion red giant star that it undergoes a thermonuclear explosion on its surface as the gas accumulated grows so dense and hot. Though the explosion does not destroy the white dwarf, it instigates a sudden burst in brightness that can be seen from Earth, fading over a period of weeks through to several months. Van Maanen noted that one nova he had observed outshone the entirety of the Andromeda spiral. If the spiral were truly extragalactic, then van Maanen believed the energy output of the nova would have to be impossibly great for it to shine so bright at such a great distance. This was, of course, before physicists had split the atom on

RIGHT Astronomer Edwin Hubble used the 100in (2.5m) Hooker Telescope, located at the Mount Wilson Observatory in Los Angeles County, California to reveal that there are other galaxies beyond the confines of the Milky Way.
(Ken Spencer)

Earth, and before the process of novae were fully understood.

On the other side of the debate was Curtis, who believed that the spiral nebulae were beyond our Milky Way. He also recognised that there were novae in the Andromeda spiral, but didn't seem daunted by their power output. Instead, he pointed out that there seemed to be an increased rate of novae in and around the Andromeda spiral. Why should that place be so special for novae unless it really was its own galaxy, far away? He noted the distribution of the spiral nebulae, which were spread all across the sky. If they were part of our Milky Way, wouldn't they then be mostly concentrated within the band of light that is the Milky Way? Instead there were no obvious spiral nebulae in that band.

Curtis won the Great Debate of 1920, and it turned out that van Maanen had been incorrect, his observations faulty when he claimed to see the Pinwheel rotating. In 1924, the astronomer Edwin Hubble (1889–1953) proved Curtis right and Shapley and van Maanen wrong when, by peering through the eyepiece of the 100-inch Hooker Telescope at Mount Wilson, Hubble was able to measure the distance to the spiral nebulae using special pulsating stars called Cepheid variables. We'll go into greater depth about Cepheids in the next chapter, but suffice it to say for now, watching how they pulse tells us their true luminosity, and therefore their distance can be accurately calculated based on how faint they appear to us.

All of a sudden the universe became larger than astronomers could have ever imagined. Today, we know that the observable universe is 93 billion light years, with even greater expanses of space beyond this cosmic horizon that we cannot see. No longer did our Milky Way contain everything in creation. Now, it was just one galaxy among trillions. We were no longer at the centre of the universe.

From planets to stars: developing the laws of motion

The realisation that the Earth was not at the centre of the universe came slowly for some. The revolution had begun in 1543, when, on his deathbed, Nicolaus Copernicus (1473–1543) published his heliocentric (Sun-centred) model of the universe, in which Earth orbited the Sun and not the other way around. For a long time it was still considered to be heresy to believe in the heliocentric model, while some philosophers of science wanted to put their own stamp on it.

One such man was Tycho Brahe (1546–1601), who was a Danish nobleman and observer of the night sky during the late 15th century, before the invention of the telescope. Tycho had his own cosmological theory, which positioned Earth at the centre of all creation, just as Ptolemy had argued over 1,500 years earlier, but also had the Sun to one side, orbiting Earth but surrounded by the other planets, which orbited the Sun. By melding Ptolemy's model with Copernicus's, he hoped to find the perfect explanation of the cosmos and become the greatest astronomer – nay, the greatest scientist – whoever lived.

Yet Tycho had a problem, and that was his German assistant, Johannes Kepler (1571–1630). Tycho was fearful; he was aware of Kepler's talent and was afraid that the young, bright astronomer may eclipse him to reign supreme as the top astronomer of the time. Kepler, you see, was a follower of Copernicanism, and he didn't just set out to follow Copernicus' lead – he was intent on improving upon it.

Tycho became somewhat paranoid; if he fired his assistant he could tip Kepler's hat that Tycho was on to something, but he was equally unwilling to share this information with Kepler. Tycho therefore had a plan. He would keep Kepler busy with a small proportion of the reams and reams of data he had collected during observations of the planet Mars. Without a complete picture, he would instruct his assistant to figure out and fully understand the orbit of the Martian world; a move that he thought would keep Kepler busy while he worked tirelessly on his theory of the solar system.

Because Tycho was exceptionally protective

RIGHT Brahe's work, *Astronomiae Instauratae Progymnasmata* (*Introduction to the New Astronomy*). *(Tycho Brahe/SLUB)*

FAR RIGHT A monument of Tycho Brahe and Johannes Kepler in Prague. *(Josef Vajce)*

of his data, Kepler at first found it incredibly difficult to deduce any real meaning from what he was given. He needed more in order to formulate what would become his three laws of planetary motion that gave mathematical structure to the Copernican model.

And then, suddenly, Tycho passed away in what were considered to be suspicious circumstances. Kepler requested access to Tycho's full sets of data, but that became difficult when the surviving Brahe family members believed that monetary gain from the deceased's observations were to be had. Due to the circumstances of his death, Kepler became a suspect in the sudden passing of Tycho. Had he poisoned him in order to get his hands on the data?

The story goes that Kepler had watched in interest as his former supervisor refused to leave a banquet in Prague, citing a breach of etiquette if he had done so despite his illness, which at the time was thought to have been a kidney stone. In his discomfort, Tycho became delirious, urging Kepler to complete Tycho's work on his erroneous model of the solar system. Tycho died on 24 October 1601, several days after falling ill, believing right up until the end that the Earth was at the centre of our universe.

Ultimately armed with the data that he needed, the student became the master and Kepler became the one thing that Tycho had feared: a premiere scientist. And it was all thanks to Tycho's data that resulted in the laws of planetary motion.

What Kepler found amazed him. Indeed, for a long time he tried to avoid his eventual conclusion, because it flew in the face of what previous, highly respected observers had insisted before him: that the planets move on a perfectly circular orbit around their parent star. Kepler's findings had revealed otherwise. Instead, that round path was somewhat flattened, squashed into an ellipse, meaning that at some points a planet was closer to the Sun in its orbit than at other points. What's more, adding irony to the story, was that the 'Red Planet' Mars swings around our Sun on a path that is among the most elliptical of all of the planets – of the planetary bodies in the solar system, only Mercury has a more elliptically

WAS JOHANNES KEPLER INVOLVED IN THE DEATH OF TYCHO?

Kepler remained a prime candidate long beyond his own death in 1630. It wasn't until 2018 that we got a complete picture of what had caused Tycho to meet his fate. His grave was opened in 1901 in an effort to get to the bottom of the mystery; there it was found that he had worn a prosthetic nose, fashioned from brass, which resulted from a drunken sword duel with his third cousin, who went by the name of Manderup Parsberg, that saw the bridge of his nose sliced off, as well as an extensive scar slashed across his forehead. Clearly, Tycho was adept at making enemies, but the subsequent re-opening of his grave in 2010 and subsequent eight-year analysis of his remains found the true cause of his death: although some traces of arsenic and mercury were discovered in his body, it turns out that his ultimate doom was his lifestyle: the results showed that a fatal dose of diabetes, alcoholism and obesity had ended Tycho's life, who claimed by his own admission that he had 'lived like a sage and died like a fool'.

shaped orbit (so too does Pluto, but it was reclassified from a planet to a dwarf planet in 2006). Tycho had unwittingly revealed to Kepler the information that would unravel and disprove Tycho's botched theory of the solar system. The Sun was at the centre of it, rather than the Earth

LEFT Tycho Brahe is buried close to the Prague Astronomical Clock, in the Church of Our Lady before Týn.
(Steve Collis)

as he'd so aggressively believed. Of course, he wasn't around anymore to learn of this.

Kepler's three laws of planetary motion are now considered to be legendary. His first law, as we have already discovered, was that all the planets must orbit around our nearest star on an elliptical path. Armed with this knowledge, he considered the speed of these orbiting worlds for his second law. You'll remember the mention of the observations that Tycho had made of Mars, which he then passed on to Kepler. From these, Kepler worked out that the Red Planet speeds up the closer it gets to the Sun, and then slows down again the further it moves away on its elliptical path. He then imagined that the two bodies – Mars and the Sun – were connected by a line made from an elastic string.

As you would expect, this imaginary line connecting them scrunches up the closer the pair get. When Mars moves further away from the Sun on its ellipse, it slows down since the elastic is stretched to an exaggerated degree. If we can picture looking down on the orbit, we would see this elastic string sweeping out equal areas over equal periods of time, no matter where the world is in its orbit around our Sun. This is Kepler's second law of planetary motion.

These two laws tell us a lot about a planet's orbital shape, allowing the orbital speed of Mars – and the other terrestrial and gaseous worlds in our solar system – to be calculated. But Kepler wasn't done: he was yet to add the finishing touch with his third law. The motions of the planets confused and bedazzled him in equal measure, but one question remained for him: why were their paths spaced out? Did they move in harmony with each other and with our Sun?

The answer didn't come to him immediately. But his persistence was rewarded by 1618, when he figured that the square of the time it takes a planet to make its way around its star – also known as its orbital period – is proportional to the cube of its average distance from its glowing host. What this essentially means is that if you timed how long it takes for Mars to complete one lap around the Sun, you'd be able to work out how far away it is.

Now we see how Kepler's discovery relates to the Milky Way at large. Just as the planets in our solar system orbit at an increasingly slower pace the further they are from the Sun, which is the centre of gravity in the solar system, stars also orbit more slowly, on elliptical paths, the further away they are from the centre of the galaxy. It should be noted that technically this is not always strictly true because of the presence of dark matter – which is a mysterious form of matter made from an invisible and currently unknown particle that interacts with normal matter only through the force of gravity – on the outskirts of galaxies. The gravity of the dark matter allows the stars to orbit faster than would be expected, but the principle of objects orbiting more slowly the further away they are from the centre of gravity in a system still stands – it was the surprising deviation from this rule that led to the discovery of evidence for the existence of dark matter in the first place.

We don't know if Kepler ever wondered to ask why this happens, but one man did, and his discovery changed everything, and provided the one missing ingredient that can explain the existence of not just the Milky Way, but all of the galaxies in the universe: gravity.

Isaac Newton's gravity

The 17th century English mathematician and physicist Isaac Newton (1642–1727) made crucial developments in numerous areas of science. He furthered our understanding of the nature of light and created a new design of telescope that used mirrors to reflect light to a focus. He discovered that a prism can split white light into a spectrum of colours – the origins of the electromagnetic spectrum – and invented the mathematical language of calculus. What he is most famous for, however, are his laws of motion and his theory of gravity.

In 1687 Newton published *Philosophiæ Naturalis Principia Mathematica*, or *Mathematical Principles of Natural Philosophy*. He'd been musing on Kepler's laws, thinking beyond the motion of the worlds in the solar system and of the behaviour of all objects in existence, from the swinging of the Moon around the Earth to an apple falling from the tree. 'To the same natural effects,' he penned, 'we must, as far as possible, assign the same causes'. In essence, this means whatever force was at play must exist everywhere; from the ground we walk on to beyond the confines of our planet and into space.

Newton spent much time observing the world around him. He noted that an object will stay put unless some kind of force acts on it. He discovered that the same can be said for anything travelling in a straight line – it won't change direction or even speed unless an outside force acts on it. Today, we recognise the phenomena as inertia; the tendency for an object to do nothing or remain unchanged unless something forces it to do so. The physicist made this his first law of motion. An example is if you blasted off into outer space, donned a spacesuit and kicked a ball into the darkness, you'd discover that it would keep travelling forever – without friction or any kind of resistance against it, there's nothing to stop it. That is, unless a meteorite smashes into it or it crashes into the gravity field of another world.

His second law is perhaps one of Newton's most familiar: F=ma. This essentially means that an object's speed will change if, again, it is exposed to an external force. The rule says that the force (F) is equal to an alteration in momentum, the latter of which can be expressed as mass (m) multiplied by velocity, over a certain period of time. Of course, when the mass remains unchanged, it's much simpler to say that velocity over time is acceleration (a). The law works both ways; a force will not only cause a change in speed, but a mass that either speeds up or slows down will also generate a force. If you've found that pushing an empty shopping trolley is much easier than manoeuvring one filled to the brim with groceries, then you've experienced this law in action; of course, more force is needed to push the cart with more mass.

'If you press a stone with your finger, the finger is also pressed by the stone', Newton stated. This is what prompted him to formulate his final and third law; to every action there is an equal, yet opposite, reaction. It explains why a diving board springs back when you jump off it, throwing you into the air and why, when you let go of an untied balloon filled with air, it whizzes around the room.

Newton had seen his laws in action on Earth, but he wasn't satisfied. He suspected that

THE MOMENT NEWTON DISCOVERED GRAVITY

Sitting in an orchard, packed with apple trees in the grounds of his childhood home at Woolsthorpe Manor, near Grantham, England in 1642, Newton had returned home from Cambridge University after an outbreak of the bubonic plague had forced his institution to temporarily close. With his back against a tree in the sunshine, Newton saw an apple fall from one of the branches and hit the ground. He shared his story with an English antiquarian William Stukeley, who later became Newton's biographer. 'After dinner, the weather being warm, we went into the garden, and drank tea under the shade of some apple trees', Stukeley wrote. 'He told me, he was just in the same situation, as when formerly, the notion of gravitation came into his mind... occasion'd by the fall of an apple, as he sat in a contemplative mood.'

they would also – in part – apply to the planets that make up our solar system. Knowing that a world would travel in a straight line unless otherwise caused not to do so by a force, he mused over the findings of Kepler, who had revealed that a planet orbits the Sun in an ellipse. There must be a force there, one that was yet to be identified.

Gravity was the answer. For Newton, it all seemed to make sense; together his laws of motion and the newly pronounced force explained Earth's yearly journey around the Sun. The physicist had been inspired to develop a law of universal gravitation; a rule that says that every little particle in the universe is attracted to every other. In the orchard that day, Newton had witnessed the true laws of nature truly in motion: he'd discovered that the force that caused the apple's acceleration in its fall, which we now know to be gravity, must be wholly reliant on the mass of the Earth. This led the physicist to conclude that the gravitational force acting between our planet and say, the Sun, is directly proportional to the Earth's mass, proportional to the mass of its orbital partner and inversely proportional to the square of the distance that separates them. Essentially, this means that gravity weakens the further you move away from an object; the planets in our solar system experience the force at varying degrees thanks to a mix between their differing masses and orbits – whether they're nearby or far flung.

ABOVE A 'bone' of the Milky Way runs horizontally along this image and is more than 300 light years long. This infrared image makes it appear dark, although it contains roughly 100,000 suns' worth of material. *(NASA/JPL/SSC)*

BELOW Star-forming region LH 95 in the Milky Way's satellite galaxy, the Large Magellanic Cloud. *(NASA, ESA and the Hubble Heritage Team (STScI/ AURA)-ESA/Hubble Collaboration)*

The gravitational universe

Gravity is the fundamental force that determines the formation and development of galaxies like our Milky Way. When our Milky Way galaxy formed, some 12 to 13 billion years ago, it was gravity that attracted enormous amounts of matter to accumulate and coalesce to form the basis of our galaxy. It's gravity that causes gas clouds to collapse and condense so densely that nuclear fusion reactions ignite in the core of the collapsing cloud, turning it into a star. It's gravity that then seals the fate of massive stars when they reach the end of their lives and their nuclear reactions stop, causing the core of the star to collapse, the resulting shockwave producing a supernova that rips the star apart and blows vital heavy elements all across space. It's gravity that hints at the existence of dark matter, which there is so much of that it accounts for 95 per cent of the entire mass of the Milky Way, forming the superstructure inside which the Milky Way was made. And it's gravity – this time of the more sophisticated flavour developed by Albert Einstein in his general theory of relativity – that dominates the physics of black holes, which have gravity so strong that they can pull in all kinds of matter, from gas cloud to asteroids and from stars to planets, and not even light can escape their gravitational pull. So-called supermassive black holes, which have masses that are millions and sometimes even billions of times greater than the mass of our Sun, are found at the cores of most large galaxies, and through some mechanism that astronomers still don't fully understand, the mass of the black hole correlates to the mass of the bulge of a galaxy, and the stopping of star formation in a galaxy via some mode of energetic feedback. But ultimately, without gravity, none of this would happen.

The black hole aside, the gravitational physics of the Milky Way is described by Newton's theory of gravity. Einstein's theory of gravity, which advances upon, rather than replaces, Newton's theory, comes into its own more in strong gravitational fields. However, the gravitational acceleration of stars and galaxies orbiting the Milky Way is low, so Newton's gravity works just fine in this circumstance.

Galaxy zoo

It's a galactic jungle out there, and the Milky Way is just one galaxy in a universe crammed with hundreds of billions of galaxies.

Broadly speaking, you'll remember that we know the Milky Way to be a barred spiral galaxy, a structure that possesses a feature akin to a wrench holding the galaxy's arms together. As spiral galaxies go, a bar is quite a common feature to have – two-thirds of spirals have them, with the remainder being of the more pure-bred variety: your normal, standard spiral galaxy.

It gets a bit more complicated when it comes to classifying this breed of cosmic structure. And attempting to file them into categories was all too familiar to Edwin Hubble, the American astronomer whom we met earlier. In 1926, he presented what we recognise today to be the Hubble Sequence, or due to its pronged appearance, the Hubble Tuning Fork. Hubble wasn't just the very first person to discover what galaxies actually are, he was the first to classify them; a scheme that has stayed in place for over 90 years.

During his observations, Hubble noticed that some of the spirals that he could see through the Hooker Telescope appeared to be more compact than others; some with their arms tightly wound while others had a much looser, more relaxed profile. He figured that a lettering system was needed – 'a' through to 'c' – where Sa represented the spiral with the tightest corkscrew arms and a bigger, brighter centre that gets smaller the looser the arms become

ABOVE The barred spiral galaxy UGC 12158 in Pegasus (the Winged Horse) is thought to resemble our Milky Way in appearance. *(ESA/Hubble & NASA)*

BELOW The spiral galaxy NGC 1232 in the constellation Eridanus (the River). Spirals account for around 60 per cent of galaxies in today's cosmos. *(ESO)*

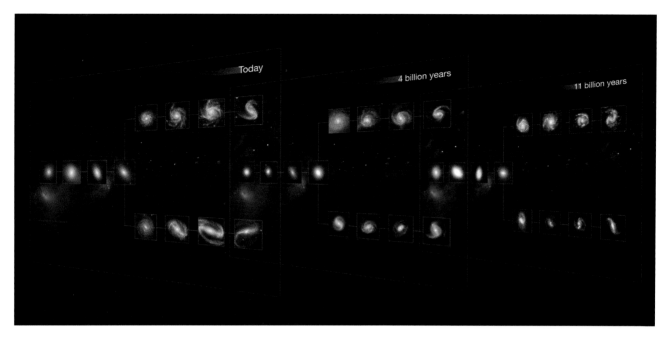

Today

4 billion years

11 billion years

ABOVE The evolution of the Hubble sequence throughout the universe's history. *(NASA, ESA, M. Kornmesser)*

(Sb and Sc). He composed the same method for the barred spirals, but with the prefix SB (i.e. SBa, SBb, SBc).

But there was more. As he slewed his mighty telescope across the night sky, Hubble came across a whole plethora of galaxies with no grand features to speak of. These were the featureless elliptical galaxies (unrelated to the elliptical orbits described by Kepler). These egg-shaped, featureless structures sit on the left of Hubble's fork and, while devoid of any features

that distinguish them individually, there's one thing that separates them into types, which is how squashed they are. Visually, some are round, while others take on a more flattened rugby-ball appearance.

Hubble recognised this immediately during his observations. As he'd done previously for the spiral galaxies, he ordered the ellipticals into how flattened they are, with E0 being the most round and E7 being the most ellipsoidal.

Unlike their cousins, the spirals, the ellipticals

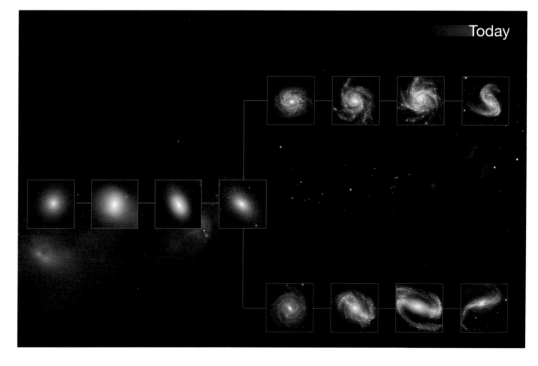

Today

RIGHT A basic version of the Hubble Tuning Fork, showing elliptical, spiral and barred spiral galaxies. *(NASA/ESA)*

are at the opposite end of the star-forming spectrum. Galaxies with arms are constantly teeming with young stars, circulating the dust and gases that make stellar formation possible. In the world of the elliptical, the environment is anything but – the conditions are barren, with literally no means of making an abundance of young, luminous stars possible. They're three-dimensional in structure, their older, more seasoned stellar members making their way around the galaxy on random orbits, hence they appear spherical rather than disc-shaped. Additionally, without bursts of stellar activity that give spiral galaxies their characteristic hot white-blue palette (massive stars formed in vibrant star-forming regions are so hot they shine blue), the elliptical is painted red, the birth of its stars a distant memory, formed in the galaxy's heyday billions of years ago, as only the smallest, coolest, longest-lasting stars survive. They are the stellar veterans of the universe.

The universe is wild, meaning that not everything fits into a perfect set up. Hubble knew this and soon realised that some of the galaxies he pored over didn't quite fit his model. Slotting the lenticular galaxies – which are featureless disc galaxies that seemed to share properties of spiral and elliptical galaxies and which he dubbed class S0 – into the picture seemed to be an easy task. They weren't too dissimilar to the elliptical, especially when Hubble observed them head-on, with their bright centres surrounded by a flattened disc. Side-on, you'd see a completely different story as they reveal lanes of dust, a feature that isn't too dissimilar to the spiral galaxy. The appearance of the lenticular seemed to flip depending on which angle you looked at them; for this reason, the class became a centrepiece in Hubble's picture of classification, linking the spirals, barred spirals and ellipticals into one complete, seamless picture.

Although Hubble did not intend it to come across this way when he drew up his tuning fork, it was this seamless picture that led some astronomers to suspect that there was an evolutionary picture emerging from the diagram. They thought that the elliptical galaxies, with their similarities to the bulges of spiral galaxies in terms of shape and star type, formed first, and then they developed spiral arms afterwards.

It turns out that they got it backwards, with spiral galaxies not necessarily being younger, but certainly undergoing continued evolution, whereas the ellipticals ended their evolution billions of years ago, and were often formed when two spiral galaxies collided and merged together. As we shall see in the final chapter, this will be the ultimate fate of the Milky Way.

However, spirals and ellipticals are not all that is out there. It was the presence of the irregulars – galaxies with no defined shape, with patches of stars mangled into a cosmic mess, possibly the result of two galaxies in the midst of smashing into each other over a period of millions of years – that caused Hubble to begin to question whether his tuning fork was as all-encompassing as he'd originally believed. That was something that French astronomer Gérard de Vaucouleurs (1918–95) was entirely sure of when, in 1959, he compared what he witnessed during his galactic tours of the sky to Hubble's Tuning Fork. So he developed a modified version of Hubble's Tuning Fork for characterising galaxies, keeping the ellipticals, spirals, barred spirals and lenticulars, but adding a lot more detail. To look at, de Vaucouleurs's

BELOW A close-up of a galaxy's spiral arm – where gas and dust is squeezed by gravity to form brand-new stars. *(NASA/ESA)*

LEFT Some galaxies are devoid of any particular shape. They are classed as irregulars, an example being NGC 1427A. *(NASA, ESA and the Hubble Heritage Team (STScI/AURA))*

version is much more extravagant, with extra classifications for spiral galaxies taking into account whether a galaxy with a bar was weak or strongly barred (and if they had no bar they were designated 'SA'), whether they had an inner ring-like structure (signified by the labels '(r)' for galaxies with rings, '(s)' for galaxies without, and '(rs)' for transitional galaxies in between being ringed and not being ringed), and their overall structure ('Sd' for diffuse, broken arms; 'Sm' for an overall irregular appearance; and 'Im' for highly irregular).

The basic Hubble Tuning Fork diagram was considered the gold standard for all galaxy classification schemes, but one criticism that astronomers had of Hubble's work was that they felt that the appearance of a galaxy was entirely in the eye of the beholder. As a case in point, how do you objectively define how tight a spiral galaxy is wound? What might have been interpreted as a loosely bound spiral to one observer, could very well have been seen as a tighter spiral to another. So de Vaucouleurs also included a numerical grading scheme to his version of the tuning fork to better define the variations in tightness of spiral arms, degree of ellipticity or prominence of the central bar.

Even so, optical observations alone don't truly give the full picture of a galaxy's structure, and Hubble and de Vaucouleurs didn't have the luxury of observing through different wavebands offered by the space- and ground-based telescopes of today. How they characterised galaxies was purely based on what they could see with the human eye, but this meant they were missing several fundamental pieces.

With the radio astronomy revolution in the

LEFT The giant elliptical ESO 325-G004, which sits at the centre of large galaxy cluster Abell S0740. *(NASA, ESA and the Hubble Heritage Team (STScI/AURA); J. Blakeslee (Washington State University))*

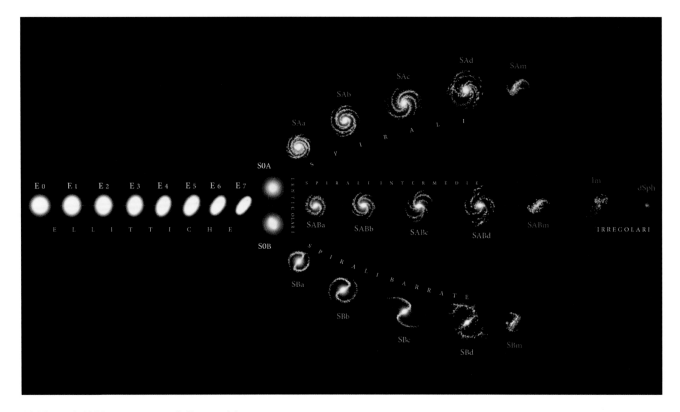

ABOVE Gérard de Vaucouleurs developed Hubble's Tuning Fork idea, incorporating galaxies that didn't quite fit into the original model. *(Antonio Ciccolella)*

1950s and 1960s, strange and distant objects with powerful radio emission began to be detected. The most famous example was 3C 273 – the 273rd object in the 3rd Cambridge Catalogue of Radio Sources. We now know 3C 273 to be the closest example of a quasar – a galaxy with an extremely active centre that is shining powerful beams of radiation towards us from the neighbourhood of a highly active supermassive black hole in the galaxy's centre. 3C 273 is located 2.3 billion light years away and it was later discovered that quasars also shine brightly in X-rays, while there is also a scale of different levels of galaxy activity, from the relatively low-level Seyfert-type spiral galaxies with bright cores to giant elliptical galaxies emanating radio jets, all the way up to the powerful quasars and the even more powerful blazars.

Another example of how looking beyond visible light in the electromagnetic spectrum can teach us about galaxy types is how star formation stands out in ultraviolet light. Intense regions of star formation in a galaxy produce a lot of hot, massive stars for which a significant proportion of their light output is at ultraviolet wavelengths. Therefore, ultraviolet light acts as a tracer for star formation, and galaxies that are in the midst of a so-called 'starburst', where their spiral arms light up with many pockets of accelerated star formation, stand out in ultraviolet light.

Of course, some of the structure seen in visible light can also be enhanced at other wavelengths. In particular, radio astronomy mapping of neutral hydrogen gas, which fills the disc of a galaxy, at the emission wavelength of 21cm can teach astronomers about the general shape of the spiral arms of the Milky Way, for example, while mid- and far-infrared light reveals the skeletal structure on interstellar dust that also defines the spiral arms. It's safe to say that if Hubble and de Vaucouleurs were alive today, they'd probably classify the galaxies in a very different manner. But, if anyone was wondering, de Vaucouleurs' classification for the Milky Way is SB(rs)bc II – in other words, a barred spiral galaxy with a transitional ring around the galactic centre and relatively loose arms. We'll find out how our Milky Way galaxy ended up having these properties in Chapter two.

Chapter Two

How the Milky Way was made

Our Milky Way galaxy is enormous, spanning 100,000 light years across and containing about 200 billion stars. It has a total mass – which includes stars, gas, dark matter and a supermassive black hole – that is 1.5 trillion times greater than the mass of our Sun (one solar mass is 2×10^{30} kilograms). Yet in the grand scheme of things it is a fairly average spiral galaxy. But, that doesn't mean that it's not still a remarkable place, with a remarkable history, one that we're learning more about all of the time.

OPPOSITE Giant molecular clouds, like the Orion Complex (pictured), are star-making factories comprised primarily of gas. They are massive, between 1,000 and 10,000,000 times the mass of the Sun and can extend for 15 to 600 light years. *(ESO, Digitized Sky Survey 2)*

Piecing together that history is not easy. Much of it happened a long time ago, billions of years before our Sun had even formed by coalescing from a giant cloud of molecular hydrogen gas. Our galaxy likely formed much like other galaxies though, so by peering deep into the universe, our telescopes are able to capture snapshots of galaxies in the act of being assembled, and then evolving further as they mature over time to become like the Milky Way and other galaxies that we see in the local universe.

That's one way of figuring out some of the details of how the Milky Way formed, by looking outside it. The other way is to look inside, and to learn to decipher the messages that the stars and structures within the Milky Way are telling us. Either way, we must first go back to the very beginning – the Big Bang – to understand the basis for everything about galaxies like the Milky Way.

The Big Bang

The Big Bang was the event that kick-started the universe. Scientists do not currently understand what caused it, but do understand the seconds and minutes that followed in the Big Bang's immediate aftermath in great detail.

At the very moment of the Big Bang, the universe was a single, dimensionless point of infinite density. Instantly it began to grow, in the process experiencing a brief – but rapid – blast of expansion termed inflation, which lasted between 10^{-36} and 10^{-32} seconds after the Big Bang and increased the size of the universe from microscopic to macroscopic at a rate faster than the speed of light. As a result, parts of the embryonic universe that had been in causal contact with each other were forever carried away.

Although inflation quickly shut down, the universe continued to expand, driven by the energy of the Big Bang. As it expanded, it cooled rapidly, allowing subatomic particles to condense out of the raw energy that filled the universe. These particles were quarks, which in turn combined to form meson particles made of two quarks, and more importantly baryons made of three quarks. These baryons are protons and neutrons – the very same particles that form atomic nuclei. Within 20 minutes of the Big Bang, the nuclei of the very first elements had formed. At this time, the universe was filled mostly with hydrogen nuclei formed from a single proton (along with its isotopes deuterium – which combines a proton with a neutron – and tritium, which contains one proton and two neutrons) amounting to about three-quarters of all the baryonic matter in the universe. Helium-4 nuclei, which is made from two protons and two neutrons, and helium-3 made from two protons and one neutron, composed most of the rest, with lithium and beryllium amounting to a small fraction of a per cent. The beryllium-7, however, promptly decayed with a half-life of 52 days into lithium-7 thanks to the weak force changing one of its protons into a neutron. So, ultimately, after the Big Bang there was just lots of hydrogen and its isotopes, some helium and its isotopes, and a very tiny amount of lithium.

For the first 379,000 years of cosmic history, this was the state of the universe, filled with a plasma of atomic nuclei and free electrons, the conditions too hot to allow the nuclei and electrons to join to form complete atoms. During the first ten seconds, the universe had created almost as much antimatter as it had matter, which promptly annihilated, leaving only the excess of matter left to form everything that

WHAT CAUSED INFLATION?

The driving force behind inflation has never been fully explained, but it occurred at a time when the fundamental forces of nature were just beginning to break away from a primordial unified force that 21st-century science is yet to understand. In particular, the energy for inflation may have been released when the nuclear strong force (which acts to hold atomic nuclei together) separated from the electroweak force, which was the combined form of electromagnetism and the weak force (which causes radioactive decay). This separation of the forces caused a kind of phase transition in the early universe – one analogy is how water evaporating from a liquid into a vapour releases latent heat.

Whatever the driving force behind cosmic inflation, we do know that it must have been an energy field that pervaded the universe. Such energy fields are subject to quantum physics, whereby Heisenberg's Uncertainty Principle results in temporary variations in the strength of that energy field. These so-called quantum fluctuations were also enlarged by inflation.

we see in the universe today (why antimatter and matter was not created in equal amounts remains one of the greatest scientific mysteries). During each annihilation, the doomed matter and antimatter produced pairs of photons of light. However, the plasma ocean that filled the universe proved to be a substantial barrier to these photons – the free electrons in the plasma constantly scattered the photons, reducing their energy. Matter and radiation was therefore said to be 'coupled' together, and light could not really travel through the universe. It was a dark, murky time.

The enlarged quantum fluctuations produced by the inflation field had not been forgotten. Now they sloshed through the plasma ocean like waves, with crests coinciding with the plasma being bunched up, and troughs where the plasma was able to spread out a little more.

Then, after 379,000 years, everything changed. The universe had cooled to about 3,000°C (5,432°F), which is cool enough for the free electrons within the plasma to combine with the atomic nuclei to form complete atoms. Suddenly, without the electrons to scatter them any longer, photons could travel through space unhindered. Cosmic expansion across 13.8 billion years has since stretched this light to microwave wavelengths, and we can observe it today as the cosmic microwave background (CMB) radiation. Predicted as a consequence of the Big Bang and the expansion of the universe by cosmologists Ralph Alpher and Robert Herman in 1948, the CMB was famously discovered quite by accident in 1964 by Arno Penzias and Robert Wilson at Bell Labs in New Jersey, United States. It reflects the 'moment of last scattering', in other words, how the universe appeared the last time those photons scattered off an electron.

With the plasma gone, the giant waves became frozen in place, their crests remaining as regions of slightly more dense matter compared to the troughs. These frozen waves manifested themselves in the CMB as regions of slightly greater or lesser temperature, termed anisotropies (meaning they are direction dependent, as opposed to isotropic, which refers to a property being the same in all directions). These anisotropies were first detected by NASA's Cosmic Background Explorer (COBE) and revealed to the world in 1992, an achievement that won COBE's two principal investigators, George Smoot and John Mather, the 2006 Nobel Prize in Physics. It certainly was an impressive achievement: the variations in temperature in the anisotropies are incredibly difficult to discern, varying by only one part in 100,000 relative to the average temperature of the CMB, which is minus

BELOW The cosmic microwave background, the relic radiation from the birth of the universe, as seen by the ESA Planck satellite in 2013. *(ESA)*

270.4°C (minus 454.7°F). NASA's follow-up mission, the Wilkinson Microwave Anisotropy Probe (WMAP), as well as the European Space Agency (ESA)'s Planck mission, have been able to show these anisotropies in increasingly greater detail.

The importance of the anisotropies is that they represent regions of denser matter in the early universe, regions that would act as the seeds upon which galaxies, including the Milky Way, would form. It's an incredible connection between the large-scale structure in the universe today, and the quantum fluctuations that occurred during the first tiny fractions of a second after the Big Bang.

The denser regions at the moment of last scattering contained a little bit more matter than the less dense regions, and their stronger gravity was enough to attract more matter,

pulling it away from the less dense regions. As the universe continued to expand, both the less dense and more dense regions expanded too. Today, the denser anisotropies have developed into the cosmic web – a universe-spanning network of filaments of matter, with the nodes of the network being where galaxies and galaxy clusters are concentrated. Meanwhile, the less dense regions expanded to become cosmic voids with a more sparse population of galaxies.

A hierarchy of galaxies

The visible baryonic matter in the universe, which makes up all the planets and stars and interstellar gas that we can see in the cosmos, actually amounts to only 4.9% of all the matter and energy in the universe. It's an astounding statistic that means the vast

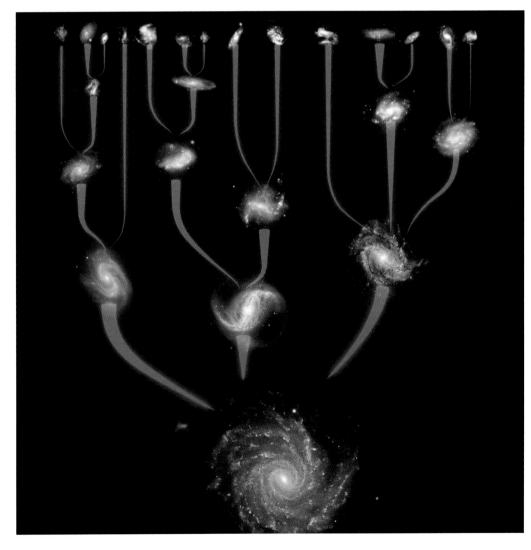

RIGHT The hierarchical model of galaxy formation states that small galactic building blocks merge to form into larger structures throughout the lifetime of the cosmos.
(ESO/L. Calçada)

majority of the universe is unseen and unknown. Some 68.3% of the universe is dark energy, which is the mysterious energy field that is causing the expansion of space to accelerate. More pertinent to understanding how galaxies, including the Milky Way, formed is the remaining 26.8%, which consists of dark matter.

Because there is so much dark matter, it fills the cosmic web, and forms massive haloes that attract more normal matter with their gravity. Indeed, our Milky Way galaxy is ensconced within an enormous halo of dark matter containing 95 per cent of our galaxy's total mass, equivalent to up to three trillion times the mass of the Sun.

Nobody knows what dark matter is, but the current prevailing model is that it is 'cold', i.e. of low temperature and low energy, causing dark matter particles to move sluggishly and rarely, if ever, interact with normal matter except through

ABOVE The network of filaments of dark matter – known as the cosmic web – in which galaxies are believed to be embedded and form. *(AVL NCSA/ University of Illinois)*

BELOW The Milky Way features a thin disc, which contains stars with a wide range of ages, and a thick disc of energetic stars that forms from the heating of the thin disc. *(ESO/NASA/JPL-Caltech/M. Kommesser/R. Hurt)*

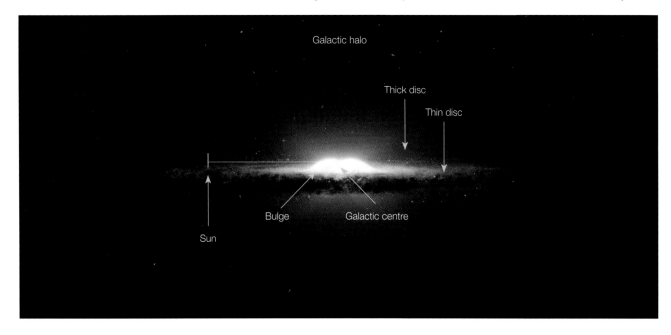

Galactic halo

Thick disc

Thin disc

Bulge

Galactic centre

Sun

ABOVE An artist's impression of a dark matter halo around the Milky Way galaxy. *(ESO/L. Calçada)*

gravity. Under this paradigm, galaxies are built from the bottom up. Baryonic matter flows into the dark matter haloes that are concentrated at the nodes of the cosmic web, and that baryonic matter, which is in the form of streams of mostly hydrogen gas, form massive gas clouds that under self-gravity collapse to produce small galaxies known as dwarf galaxies. Typically these dwarf galaxies are a few thousand light years across and contain a few hundred million, to a few billion, solar masses. In what is described as a hierarchical process, these dwarf galaxies sink to the centre of the dark matter halo, where they collide and merge, gradually building up into a larger, roughly spherically shaped galaxy. This is how the proto-Milky Way would have looked around 12 to 13 billion years ago.

The picture of hierarchical formation as described above is, admittedly, a simplified picture of how large galaxies form – observations of the early universe made by the Hubble Space Telescope, for example, have revealed a more complex interplay of numerous flows of gas streaming onto young galaxies, influenced by the gravity of the dark matter haloes, versus feedback effects resulting from the X-ray and ultraviolet radiation from young massive stars that form in the galactic maelstrom, as well as from powerful winds blowing from hot discs of gas circling the growing black holes at the centre of these young galaxies. However, simplified though it is, the hierarchical model gives us the general picture.

However, this leads to a modern-day problem. The paradigm of hierarchical formation

inside haloes of cold dark matter predicts that there should be plenty of leftover dwarf galaxies still in the vicinity of the Milky Way – as many as between 200 and 500, depending on whose computer simulation you believe. However, less than 60 dwarf galaxies have currently been identified out to a distance of 1.4 million light years from the Milky Way, and not all of them are gravitationally bound to our galaxy. This doesn't necessarily mean that the hierarchical formation model is in trouble. Some of the recent dwarf galaxy discoveries, such as that of the Virgo I dwarf, which is 280,000 light years away and was discovered by the 8.2-metre Subaru Telescope in Hawaii, are incredibly faint. These so-called 'ultra-faint dwarfs' contain a greater abundance of dark matter than the brighter dwarfs, and fewer stars, making them difficult for our telescopes to find. Therefore, it could well be that most dwarf galaxies are dark matter-rich and star-poor, rendering them so faint that they will require deep and careful surveys of the heavens to find them.

Building the disc

Of course, the Milky Way is not spherical, as the proto-Milky Way would have been, but is predominantly disc-shaped – indeed, the spiral disc is perhaps our galaxy's most defining characteristic. However, with all those dwarf galaxies colliding and merging from all angles, and ditto the streams of gas accreting onto the young galaxy, the spherical proto-Milky Way began rotating. As it grew in mass via all these minor mergers (galaxy mergers between a larger galaxy and a small galaxy are referred to as 'minor', whereas mergers between two large galaxies are described as 'major'), it began to rotate faster and faster, and in order to conserve angular momentum, much of the gas in the nascent galaxy flattened out into a disc that was spinning perpendicular to the axis of rotation.

This flattening produced what we refer to as the 'thick disc', which is about 1,000 light years thick. It is filled with stars older than ten billion years, indicating when the thick disc must have formed. Today, however, the thick disc is inactive, having used up all its gas for forming new stars. Instead, most of the star-formation action takes place today in the thin disc, which is embedded within the thick disc and is about 400 light years thick, on average. The thin disc formed after the thick disc, less than 10 billion years ago, and contains four-fifths of all the visible matter in the Milky Way galaxy. It is effectively what we see when we look up at the sky on a dark night and witness the shimmering stream of the Milky Way arcing across the sky. The reason that the thick disc is, well, thicker, is because when it formed, the orbits of its stars were more energetic than those formed in the thin disc, and therefore the thick disc is more 'puffed up'.

In the context of the assembly of the Milky Way, whereas the thick disc may have formed by the accretion and cannibalisation of dwarf galaxies falling onto the Milky Way from the cosmic web, the thin disc formed via a gentler process called 'secular evolution'.

Stars form from clouds of molecular hydrogen gas. In the Milky Way, and other spiral galaxies, there are the giant molecular clouds (GMCs) – vast, lumbering agglomerations of gas, which can reach up to 300 light years across and have masses 100,000 times greater than the mass of our Sun. However, a molecular cloud of such stature doesn't simply give birth to stars unprompted: it requires something to disturb it, causing the gas within the cloud to become dynamically unstable to the point of fragmentation and gravitational collapse. The

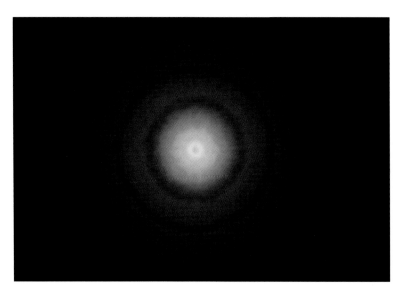

interaction of a dwarf galaxy falling through and merging with the Milky Way galaxy would be one way to disturb the GMCs, but such events are infrequent and localised in our galaxy.

On closer inspection, we find that a lot of the star formation in the Milky Way's disc takes place in the spiral arms. The Milky Way has two primary spiral arms – the Scutum–Centaurus Arm that winds around the far side of the Galaxy, and the Perseus Arm. Two further arms, named the Norma Arm and the Sagittarius spiral arm, are relatively minor. Our solar system resides in a small spur called the Orion Arm, which is interior to the Perseus Arm.

ABOVE This image of the TW Hydrae protoplanetary disc is the best to date. Population III stars would have kick-started the formation of planets. *(S. Andrews (Harvard-Smithsonian CfA); B. Saxton (NRAO/ AUI/NSF); ALMA (ESO/ NAOJ/NRAO))*

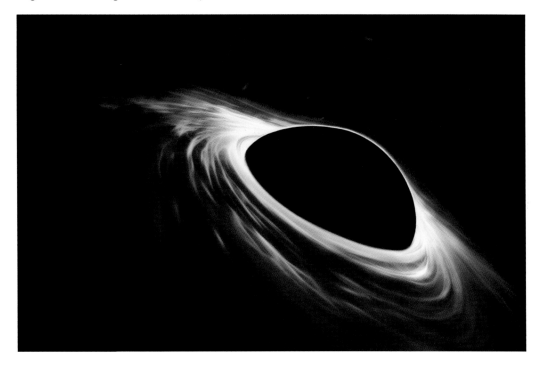

LEFT The supermassive black hole at the centre of the Milky Way is spinning, surrounded by a disc of rotating gas dubbed an accretion disc. *(ESO, ESA/Hubble, M. Kommesser)*

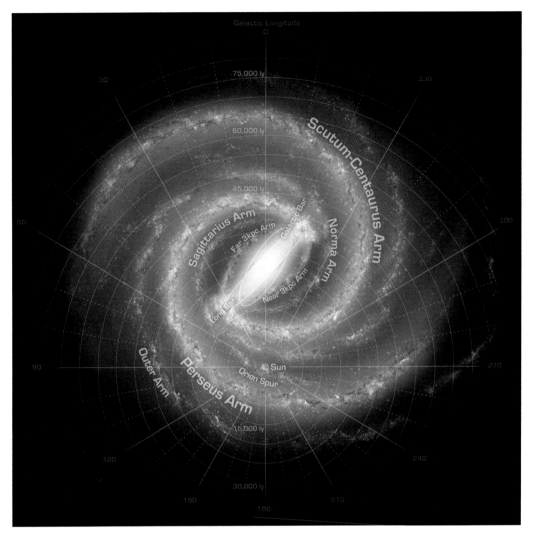

The thing about the spiral arms is that their contents are constantly changing – the Sun, for example, won't be a member of the Orion Arm forever. Stars and gas closer to the centre of the galaxy rotate faster than stars at larger radii – the simple consequence of Kepler's third law of orbital motion. This variation in velocity is known as differential rotation. If the stars remained entrenched within their spiral arms – i.e. the arms were rotating at the same velocity as the stars and gas within them, so that the same stars and gas always stayed inside their spiral arms – then this differential rotation would very quickly cause the spiral arms to wind up tighter and tighter around the galaxy. In fact, given that the Sun takes about 225 million years to orbit the galaxy, then that typical rotation period should have seen the arms wind around the galaxy 50 or 60 times by now.

Clearly this winding does not happen, in the Milky Way or any other spiral galaxy that we see in the universe. It became known as the Winding Dilemma, but one credible solution was developed in the 1960s by Chia-Chiao (C. C.) Lin of the Massachusetts Institute of Technology. He calculated that any unevenness in the distribution of matter across the rotating disc of a galaxy would lead to instability, creating two large, spiralling density waves where stars and gas are bunched up. The waves themselves are not a physical object; instead, think of a traffic jam on the motorway. Cars approach and catch up to the traffic jam, then slow down while they are in it, before speeding up again when they reach the front of the queue of traffic. In much the same way, Lin described how stars orbit the galaxy faster than the motion of the density waves, meaning they catch up to the trailing end of the waves, progress through them more slowly, before

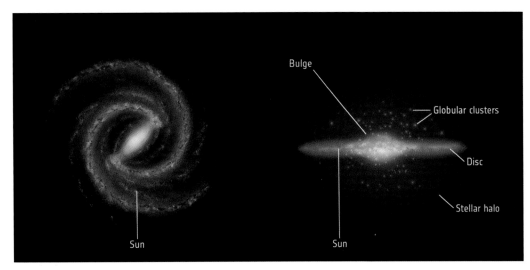

Bulge

Globular clusters

Disc

Stellar halo

Sun

Sun

LEFT An artist's impression of our Milky Way galaxy, displayed face- and edge-on. The side view reveals the bulge, disc, halo and globular clusters. *(Left: NASA/JPL-Caltech; Right: ESA)*

speeding up again when they get out of the other side.

When the stars pass through the density waves, nothing happens to them, but when the enormous GMCs do, the sudden change in their overall velocity causes a supersonic shockwave to ripple through them, the same way that a car might crumple when it crashes into a wall. The shockwave causes the gas in the GMC to compress, collapse and fragment, creating dense nebulae that form stars. That's why in the Milky Way and other spiral galaxies, we see a lot of star formation in the spiral arms.

The galactic bar and bulge

The Milky Way's disc also presents another distinguishing feature, specifically a central bar. The bar is also a product of secular evolution, whereby instabilities in the disc, caused by the disc's own varying gravitational fields produced, for example, by the massive GMCs, causes the inner disc to buckle and the bar to form.

The bar runs through the centre of the Milky Way, linking the two primary spiral arms – the aforementioned Scutum–Centaurus Arm and the Perseus Arm – that protrude from either end of the bar. The Milky Way's bar is large, although its exact extent has not been properly determined; estimates place it at between 3,000 and 15,000 light years in length, extending either side of the centre of the galaxy. Measuring the dimensions of the bar is complicated by several factors, such as it being

hidden behind the gas and dust in the plane of the Milky Way's thin disc, and the fact that we seem to be looking more down the length of the bar from our vantage point in the galaxy; measurements suggest that the end of the bar is pointed almost directly towards us.

It's not really surprising that the Milky Way has a central bar. Actually, surveys of galaxies suggest that about two-thirds of spiral galaxies have central bars made of stars and gas, and the suspicion is that at one time or another, most spiral galaxies develop bars, which implies they may be a transitory feature, coming and going on timescales of billions of years.

The Bulge Radial Velocity Assay (BRAVA)

BELOW The Laser Interferometer Space Antenna (LISA) is designed to hunt for ripples in space-time known as gravitational waves from our galaxy, proven to be the result of merging black holes among other energetic activity. *(NASA)*

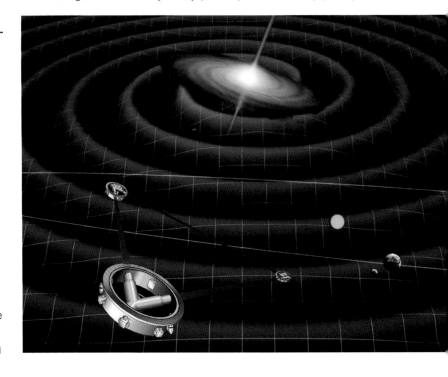

project, using the 4m (13ft 2in) Blanco telescope at the Cerro Tololo Inter-American Observatory (CTIO) in Chile, spent four years collecting spectra on 10,000 red giant stars within the bar of the Milky Way. In 2011 astronomers, led by Michael Rich of the University of California, Los Angeles, released the results of the BRAVA project, which shed new light on the origin of the bar. The measurement of the Doppler shift – the stretching and compression of the wavelength of the starlight depending on whether the star is moving towards us or away from us – in the spectra of light from each red giant star reveals their motion. When the information on all the motions of the stars are combined, we can learn about the overall motion of the bar. What Rich's team confirmed was that, unlike the galaxy's spiral arms that seem to be density waves of stars passing through, the bar is rotating like a solid object around its long axis, like a cylinder rotating.

The bar formed following a 'disc instability'. The idea is that when the Milky Way's disc first formed, its gravitational field was axisymmetric. In other words, the gravitational field was symmetric about the galaxy's axis of rotation. However, as we know, the disc contains vast clouds of molecular gas, spiralling density waves, and is also perturbed by the gravitational tides of neighbouring dwarf galaxies such as the Small and Large Magellanic Clouds. These conspire to turn the axisymmetric gravitational field non-axisymmetric, disturbing the path of gas clouds around the galaxy.

Kepler's laws tell us that orbiting bodies move in ellipses, and in the case of stars around the centre of the galaxy, some of their orbits can be highly elliptical. However, in an axisymmetric situation, gas clouds settle into circular orbits because of their relatively frequent collisions or gravitational perturbations with other gas clouds, which saps their orbital momentum, causing their orbits to circularise.

The introduction of these perturbations altered the dynamics of the rotating disc. It began to crumple in its centre and the overall gravitational field became non-axisymmetric, changing the orbits of some of the stars and especially the GMCs closest to the bar so that they began to flow inwards on non-circular orbits, therefore feeding the bar with material.

There are several resonances between the bar and the disc. Two of these resonances are named after the Swedish astronomer Bertil Lindblad (1895–1965), who was the first to describe them mathematically. The inner Lindblad resonance (ILR) occurs when a star (or gas cloud) is orbiting on a resonant orbit faster than the spiral arm that it is passing through, and the resonance increases the orbital velocity of the star (or gas cloud), causing the star's (or gas cloud's) orbit to widen. Kepler's laws of orbital motion describe how objects closer to the centre of mass in a system orbit faster, meaning that it is stars within the inner disc of the Milky Way that will be orbiting that much faster than the movement of the density wave spiral arm through which they are passing.

Meanwhile, the outer Lindblad resonance (OLR) is the opposite, describing a star (or gas cloud) moving more slowly on an outer resonant orbit than the section of the spiral arm it is moving through. This has the effect of slowing the star (or gas cloud), causing its orbit to shrink. Between the two Lindblad resonances is the co-rotation resonance, where a star or gas cloud is moving at the same angular velocity as the bar.

Now, orbits in the bar are found to be between the IRL and the co-rotation resonance, and it is these resonances that push stars and gas from the disc and into the bar. This

GETTING INTO THE SWING OF GRAVITATIONAL RESONANCES

The main driver of the continued evolution of bars and their apparent rotation are gravitational resonances. A resonance occurs when a force that is applied to an object coincides with some natural frequency. So in an everyday situation, there is a resonance when pushing a child on a swing: by applying the push at the same point in each swing, the amplitude of the arc's swing grows larger.

In the context of orbits around a galaxy, resonances occur when an object in an outer part of the disc orbits multiple integer (a full number) times more often than an inner part of the disc. So, for example, there will be a specific galactic radius where a star or a cloud of gas orbits exactly three times, for example, compared to one orbit within the bar. The exact relationship in the periodicity sees the bar and the object's orbit align, so that the object receives that regular gravitational 'push', a bit like the child on the swing.

happens because bars transfer their angular momentum to the disc beyond them in the process. The input of angular momentum into the disc helps drive the spiral pattern, and the increased angular momentum pushes stars and gas clouds outside of the co-rotation resonance towards the bar. As a side effect, the loss of angular momentum makes the bar grow bigger.

The BRAVA results showed that the bar is rotating like a solid object would. This is made possible because the orbits between the IRL and the co-rotation resonance precess at exactly the same rate. These orbits are elliptical, meaning that they are not constantly the same distance from the galactic centre. The precession of their orbits describes how the position of the closest point in their orbits (called periapsis) around the galactic centre changes upon each orbit. So by the stars and gas clouds all precessing together, the Milky Way's bar appears to be rotating as one.

The closer a star or gas cloud is to either the IRL or the co-rotation resonance, the greater the eccentricity of its orbit, but it also causes the axes of their orbit to switch over. This means that lots of objects end up on orbits that cross each other. This is fine for stars, which in the grand scheme of things are too small for there to be much chance that they could collide. For large gas clouds, however, it's another matter. The gas clouds in the bar end up colliding and compressing, before collapsing and fragmenting to form stars. This is further exacerbated when one considers that gas within the bar flows, in general, towards the centre of the galaxy, accelerating as it falls down the gravitational well so that it overshoots the centre and decelerates before reaching the other end of the bar. As it decelerates the gas starts to pile up, with more gas slamming into it from behind, and the shocks that reverberate through the gas clouds create more ideal conditions for star formation, particularly as the clouds begin to reach a bottleneck at the end of the bar. This is why, in the Milky Way and in barred spiral galaxies beyond, bars are one of the most likely regions in which to find intense star formation – a so-called 'starburst'.

This entire process of bar formation and keeping the bar fuelled with fresh material is a secular one. The BRAVA results strongly

imply this by matching the measurements to a model developed by Juntai Shen of Shanghai Observatory, which describes how a bar with the dynamics that we observe in the Milky Way's bar originates from a purely disc-like galaxy. This is incredibly revealing about the Milky Way's early history. In some spiral galaxies, central bulges are seen rising out of the plane of the spiral disc. These bulges have been described as being like mini-elliptical galaxies filled with old stars at the hubs of galaxies, and are understood to have formed from a multitude of mergers, as indicated in the hierarchical formation model. However, if the Milky Way had no bulge but just a disc of stars and gas, it could not have experienced many, if any, major mergers in the past, but just gentle gas flow accretion from the cosmic web and the occasional cannibalisation of small dwarf galaxies that get too close.

Instead, the Milky Way has what is called a 'pseudobulge' – also known as a 'boxy-', 'peanut-' or 'lens-shaped' bulge – which forms as a result of the bar rearranging the gas in the galaxy's disc, causing some of the gas and stars to flow through the bar and settle in the centre of the galaxy over a timespan of billions of years, and therefore the formation of the pseudobulge is a result of secular evolution. This connection between the formation of a bar

ABOVE Swedish astronomer Bertil Lindblad described the 'Winding Problem' – he believed that when a spiral galaxy rotates, the arms wind up tightly. *(NASA and the Hubble Heritage Team (STScI/AURA) Acknowledgement: Dr. Ron Buta (U. Alabama), Dr. Gene Byrd (U. Alabama) and Tarsh Freeman. (Bevill State Comm. College))*

and the subsequent formation of a pseudobulge was first put forward in 1993 by John Kormendy of the University of Texas at Austin.

If traditional bulges are like miniature elliptical galaxies, which contain old, red stars (known as 'Population II' stars in the vernacular – we'll explain this in more detail shortly) moving in random orbits around the centre of their respective galaxies, then pseudobulges sport characteristics that are more akin to the discs of galaxies, in that they have a flatter shape, their stars move on orbits mostly within the plane of the disc and all in the same direction, and they experience starbursts of stellar formation, meaning that they also contain young, hot blue stars as well as older, redder stars. The presence of these younger stars has been confirmed by analysis of nearly a decade of observations of the centre of the galaxy by the Hubble Space Telescope in an observation project conducted by astronomers led by Will Clarkston of the University of Michigan-Dearborn. However, his team's observations have also confused matters. They found a mixture of old stars depleted in elements heavier than hydrogen and helium (the aforementioned Population II stars) and younger stars with greater abundances of these heavier elements (known as Population I stars). That the bulge of the galaxy should be a melting pot of all different kinds of stars, just like the disc is, was to be expected. What was surprising was the kinematics of each population. Clarkston's team found that the Population I stars have more orderly, but faster moving, orbits than the Population II stars. If the pseudobulge had formed from purely secular processes, such a distinct divide between the two populations should not be possible. If there had been a significant merger that contributed to the bulge and introduced the Population I stars, then it would have disrupted the bar.

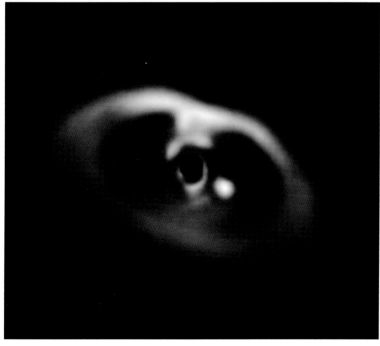

LEFT A planet is caught in the act of formation in this image from ESO's Very Large Telescope.
(ESO/A. Müller et al)

One possibility, contrary to Juntai Shen's model, is that perhaps the Milky Way did have a classical bulge prior to the formation of the bar. The formation of the bar could then have begun the process of overriding the old bulge with the new pseudobulge. Clearly, further observations are required, and it is somewhat frustrating that we can have a clearer picture of the shape and structure of galaxies millions of light years away, but because we don't have a bird's eye view of our own galaxy, the exact structure of the Milky Way remains murky.

Indeed, we're still discovering new spiral arms even now. It has been known since the 1950s that our galaxy possesses a feature called the 'Near 3kpc Arm', which is a small spiral arm about 9,800 light years (3 kiloparsecs (kpc), where one parsec is 3.26 light years) from the galactic centre and 16,900 light years from the solar system, and which arcs around the nearside of the bar. However, it was only in 2008 that a Far 3kpc Arm was discovered by Tom Dame and Patrick Thaddeus of the Harvard-Smithsonian Center for Astrophysics. The discovery of the Far 3kpc Arm provided some symmetry to the structure of the spiral disc, and in one sense the 3kpc arms could be seen as an oval-shaped ring of gas (the discovery was made at radio wavelengths using emission from carbon monoxide molecules as a tracer for the molecular gas in the arms – carbon monoxide emission is often used as a proxy for the presence of molecular hydrogen, because molecular hydrogen does not produce radio emission, unlike the 21cm line from atomic hydrogen) that seems to be expanding away from the centre of the Milky Way galaxy at a velocity of 53 kilometres per second (33 miles per second). However, it remains unclear how the 3kpc arms relate to the bar, and whether they are producing new stars from that molecular gas or not.

Stellar populations and the galactic halo

Broadly speaking, we can divide the galaxy up into two main components: the old stars and the newer stars, or in astronomical jargon, Population II and Population I stars. By identifying where the majority of each population of star resides in the galaxy, we can trace some of the milestones in the galaxy's history.

To understand the difference between Population II and Population I stars, we need to go all the way back to the early universe, in an era before the Milky Way existed. As we saw earlier in this chapter, the Big Bang produced a universe of mostly hydrogen (roughly 75%) and helium (roughly 25%), with a very tiny amount of lithium. This was all the universe had available to it to build the first stars. However, the stars are alchemists – within their interiors various nuclear fusion reactions take place that produce heavier elements – carbon, oxygen, nitrogen and so on (we'll learn more about this cosmic alchemy in Chapter 4). When a star reaches the end of its life and expires, either as an explosive supernova or the more gentle passing of a red giant shedding most of its layers, then many of these heavier elements are expelled into space, where they are eventually recycled into the next generation of stars. It stands to reason then that older stars will have fewer heavier elements than younger stars, as more and more generations of stars live and die and produce more elements.

So distinguishing stars between Population II and Population I stars is a way of determining the age of a star through its chemical content – what astronomers refer to as metallicity (the concept of referring to all elements heavier than

BELOW Infrared background light brought about by the first stars after the birth of the universe. *(NASA/JPL-Caltech/A. Kashlinsky [GSFC])*

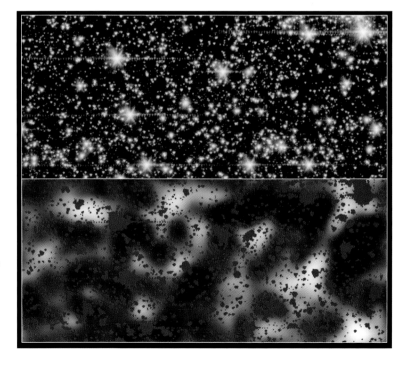

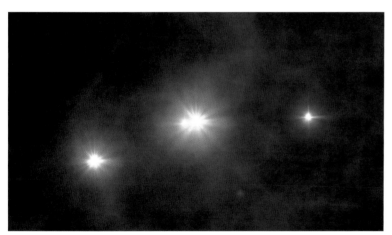

hydrogen and helium as 'metals' is unique to astronomers). Population II stars are metal-poor, i.e. they have relatively few heavy elements – between ten and 100 times less abundance than the Sun – and are thought to be over 10 billion years old. Population I stars, on the other hand, are comparatively metal-rich (though it should be pointed out that even Population I stars are mostly made of hydrogen and helium just like Population II stars) indicating a younger age of less than 10 billion years. Our Sun is a Population I star, and is 4.6 billion years old.

The German-American astronomer Walter Baade (1893–1960) was the first to identify these two stellar populations, but his most important insight was to realise that they generally inhabit different parts of our galaxy. Most of the Population I stars reside in the spiral disc, which makes sense given that we find that most of the star-forming gas in the galaxy is in the thin disc, whereas the Population II stars inhabit what astronomers refer to as the galactic halo – a huge, spherical swarm of ancient stars

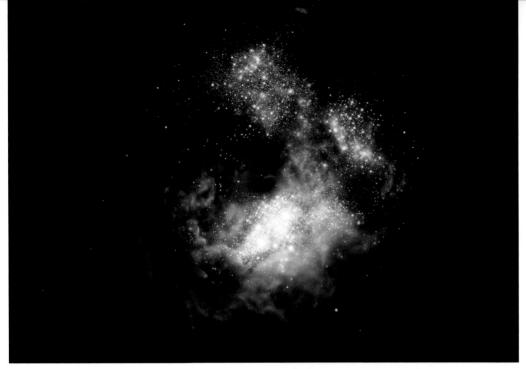

centred on the core of the Milky Way and which encompasses the spiral disc. The centre of the galaxy itself is a melting pot of stars, some fresh and young, others old. The conclusion then is clear: the Milky Way's halo formed first, with the spiral disc coming later.

Baade's system is not a perfect demarcation. The increase in the metallicity of stars was gradual over time, rather than a sudden jump that is implied by the two discrete populations. But while the correlation between metallicity, age and location holds in general for the Milky Way, it can fail for other galaxies that have strongly influenced the evolution of our galaxy – in particular, we are referring to the dwarf galaxies.

There would be no room for confusion if the stars in dwarf galaxies remained in the dwarf galaxies, but they don't. Instead, our Milky Way occasionally cannibalises dwarf galaxies that get too close, the gravitational tidal forces from our galaxy tearing them apart into long

WHAT EXACTLY ARE DWARF GALAXIES?

Dwarf galaxies come in a variety of guises – elliptical, spheroidal, irregular, even a handful with spiral formations – and tend to be chemically undeveloped compared to more massive galaxies because their star-formation rate is generally quite low, leading to lower metallicities than for stars of equivalent age born in the Milky Way.

ABOVE The Milky Way is likely to possess a central bar through its centre. *(NASA/JPL-Caltech/R. Hurt (SSC))*

BELOW The globular cluster Palomar 12, captured in this photograph by the Hubble Space Telescope, is thought to have been captured by the Sagittarius Dwarf Spheroidal Galaxy. *(NASA/WikiSky)*

streams of stars that wrap around the Milky Way. In recent decades, the Sloan Digital Sky Survey – consisting of a 2.5m (8ft 2in) telescope at Apache Point in New Mexico, USA – among other surveys, has identified about 20 faint streams of co-moving stars in the sky wrapping around our galaxy like ribbons. These include the Sagittarius Stream, which is a river of stars currently being pulled out of the Sagittarius Dwarf Spheroidal Galaxy by the gravity of the Milky Way. Over time, the stars in these streams will become fully integrated into the galactic halo.

A warped disc

One peculiar thing about the evolution of the thin disc of the Milky Way is that it is warped. If we could see it edge-on, it would appear to have a lip towards its edge, focused on the leading spiral arm. The warp was first noticed in the 1970s in 21cm wavelength radio observations of neutral (atomic) hydrogen. Several hypotheses were proposed to explain it, including interactions with the Large and Small Magellanic Clouds and the Milky Way's dark matter halo. Indeed, dark matter does seem to play some role in the formation of the warp: Albert Bosma of the Laboratoire d'Astrophysique de Marseille has found that galaxies with smaller dark matter haloes are less likely to be warped. Although we cannot see dark matter, we can detect its presence by the way its gravity bends light – or more specifically, how its mass bends space-time through which light travels, as described by Albert Einstein's General Theory of Relativity.

However, more recent studies have shone more light on the problem. In 2019, a team of astronomers led by Xiaodian Chen of the Chinese Academy of Sciences in Beijing measured the locations of motions of 1,339 Cepheid variable stars on the outskirts of the Milky Way. Cepheid variables are remarkable stars: they range in mass from four to 20 times the mass of the Sun, and are highly luminous, with their peak luminosity in some cases reaching up to 100,000 times the Sun's luminosity, which means they are visible over great distances. However, what really makes them special are their pulsations, and how

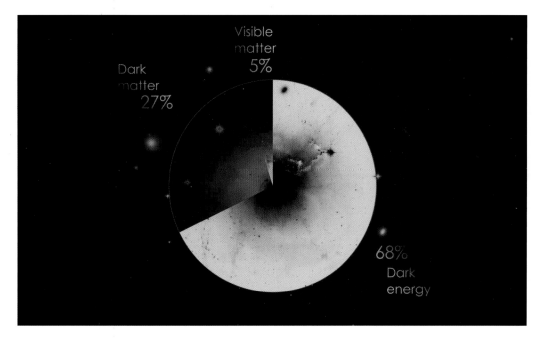

LEFT Approximate proportions of dark matter, dark energy and visible matter: the makeup of our universe. *(NASA's Goddard Space Flight Center)*

these are linked to their luminosity. As the stars pulse, their brightness changes. As first noted in 1908 by Henrietta Swan Leavitt (1868–1921) of Harvard College Observatory, the longer the period of variability, the greater the maximum luminosity of the star. Therefore, simply by measuring the period, astronomers can determine how intrinsically luminous a Cepheid variable is, and if we know its intrinsic luminosity, we can compare that with how bright it appears to us on the sky and determine how far away it must be to appear at that brightness. As such, Cepheid variables are described as 'standard candles' by which distances across the local universe can be measured.

Therefore, there are obvious advantages to observing them to determine the structure of the Milky Way, like landmarks to build a map around. Chen's group found that the positions of the stars match the warp seen at radio wavelengths, and that this warp follows the rules of warped galaxies described by Franklin Briggs, now Emeritus Professor at the Mount Stromlo Observatory of the Australian National University.

By studying other warped spiral galaxies in the universe, Briggs noted that the spiral discs are pretty much flat out to a given radius from the centre of each galaxy, and that the nodes of orbiting bodies are straight out to this distance (a node being where the orbit of an object intersects the plane of the galaxy) and precess

BELOW One of the brightest known Cepheid variable stars in the Milky Way, RS Puppis. Cepheid variables fluctuate in luminosity, allowing astronomers to use them to measure distances across the universe. *(NASA, ESA and the Hubble Heritage Team (STScI/AURA)-Hubble/Europe Collaboration. Acknowledgement: H. Bond (STScI and Penn State University))*

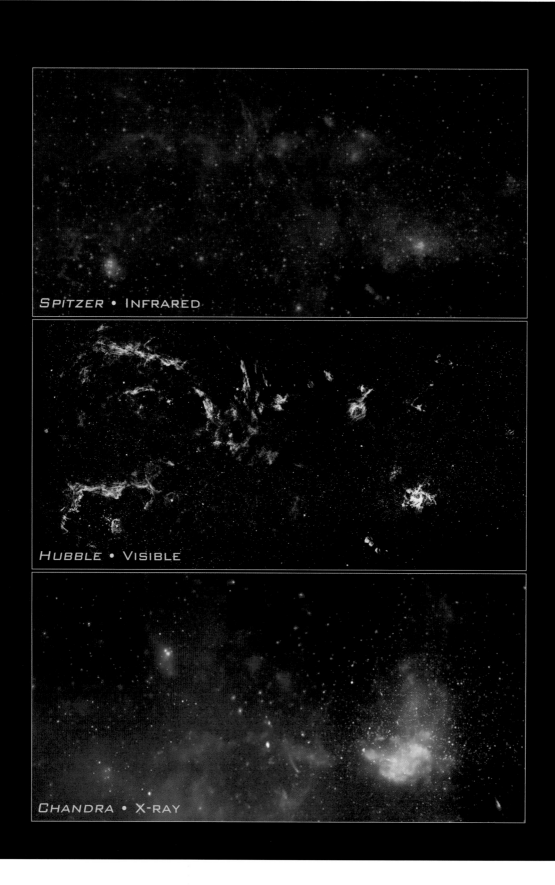

Spitzer • Infrared

Hubble • Visible

Chandra • X-ray

OPPOSITE The centre of the Milky Way visible in infrared, visible and X-ray light. *(NASA, ESA, SSC, CXC and STScI)*

around the galaxy synchronously, so that the nodes remain straight. In order to achieve this, the self-gravity of the disc must be keeping the orbits aligned. This was his first rule. His second rule was that beyond this given radius, the nodes are no longer lined up straight, but at increasingly larger radii the nodes advance around the galaxy in the direction that the galaxy is rotating. This, he says, is the result of torques acting on the outer disc by the rotation of the inner part of the disc, which contains more mass. Chen's team observed that the nodes of the 1,339 Cepheid variables matched Briggs' rules, implying that such torques, perhaps in concert with gravitational perturbations from the surrounding dark matter halo, are the guilty parties in giving the Milky Way its warp.

The stellar halo and globular clusters

We've already encountered the Milky Way galaxy's dark matter halo in this chapter, but our galaxy has another type of halo in the form of a spheroidal volume of old Population II stars extending at least 100,000 light years out from the centre of our galaxy, where the hub of the stellar halo is located.

Other spiral galaxies have haloes too, but they are relatively faint compared to the luminosity of the disc, which is why they do not appear to be obvious. Astronomers use RR Lyrae stars, which are old, pulsating yellow giants that exist in the stellar halo, as standard candles to measure distances in the halo, somewhat similar to the Cepheid variables that exist in the galactic disc. Based on their distribution, astronomers observe that the density, and therefore the luminosity, of the halo drops off with distance from the centre of the galaxy in accordance with an inverse-cubed law, i.e. $1/r^3$, where r is the radius of a star's orbit from the galactic centre. Based on this, astronomers estimate that there is 24 billion solar masses worth

ABOVE The Very Large Array (VLA) of the National Radio Astronomy Observatory (NRAO) reveals the heart of our galaxy in radio wavelengths – Sagittarius A* dazzles in radio, alongside smoke rings of supernova remnants. *(NRAO/AUI/NSF)*

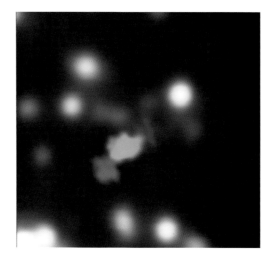

LEFT The gas cloud G2 seen being ripped apart by Sagittarius A* at the centre of our galaxy. The observations taken by the European Southern Observatory's Very Large Telescope were taken in 2006, 2010 and 2013, coloured in blue, green and red respectively. *(ESO/S. Gillessen)*

ABOVE The globular cluster Omega Centauri contains as many as 10 million stars as captured in this splendid image by the European Southern Observatory (ESO)'s La Silla Observatory. *(ESO)*

of stars in the halo – and ten times as much dark matter.

Individual stars are not the only inhabitants of the stellar halo. There are a handful of encroaching dwarf galaxies, including a handful torn apart by the gravitational tides of our galaxy, but there are also about 150 known globular clusters.

Globular clusters are great balls of stars, about 100 light years across but containing many hundreds of thousands – and in a few cases even millions – of stars. Remarkably, they are among the oldest objects in the Milky Way, having formed in gigantic bursts of star formation between 12 and 13 billion years ago. The exact process that forms such dense volumes of stars remains uncertain. At one time it had been assumed that the stars in each cluster had all formed together, at the same time. However, careful spectroscopic observations of stars in some globular clusters reveal subtle differences in the metallicity (heavy-element content) of the stars, implying that there were at least several generations of stars, with the slightly younger stars made partly out of gas expelled by the supernovae of the earlier generations. However, what appear to be new globular clusters have been observed forming around star-bursting or heavily interacting galaxies, implying that the environmental conditions play a significant factor in their formation too.

Not all globular clusters may have the same origin. Some of the largest globulars, such as Omega Centauri, contain several million solar masses of stars, and astronomers suspect that these are actually the remnant cores of dwarf galaxies that have been tidally stripped of most of their stars by the Milky Way's gravity, and the cores have become trapped in orbit. Indeed, in the globular cluster Messier 15, astronomers have detected evidence for an intermediate-mass black hole, containing about 4,000 times more mass than the Sun. Such black holes are expected to be found inside dwarf galaxies.

LEFT Hot blue stars and cooler golden stars swarm together in this shot of the globular cluster Messier 15. The collection of stars is located some 35,000 light years away. *(NASA, ESA)*

Our black hole: Sagittarius A*

If dwarf galaxies host intermediate-mass black holes (as opposed to stellar mass black holes formed by some supernova explosions), then larger galaxies such as the Milky Way host supermassive black holes, with masses amounting to millions of times greater than the mass of our Sun.

The supermassive black hole in our Milky Way galaxy is located at the galactic centre, about 25,900 light years from the solar system. There are several estimates of its mass, ranging from 4 million to 4.3 million solar masses, but whichever is the more accurate value, it is still surprisingly low for a galaxy the size of the Milky Way – for example, the Andromeda Galaxy contains a supermassive black hole that measures about 60 million solar masses.

Still, the mass of the black hole in our galaxy is still a lot to pack into its small diameter (i.e the radius of its event horizon) of about 60 million kilometres (37 million miles) across, which is slightly larger than the orbit of the planet Mercury in our solar system. Astronomers named it Sagittarius A* (A-star), which is the typical naming convention for unidentified sources of powerful radio emission. All the dust in the plane

ABOVE The centre of our galaxy, revealing a closer view of the supermassive black hole Sagittarius A*; shown as a bright radio source. *(NRAO, AUI, NSF)*

BELOW The first image of a black hole, obtained by the Event Horizon Telescope. The black hole, which is 6.5 billion times more massive than the Sun, sits at the centre of the elliptical galaxy Messier 87. *(Event Horizon Telescope Collaboration)*

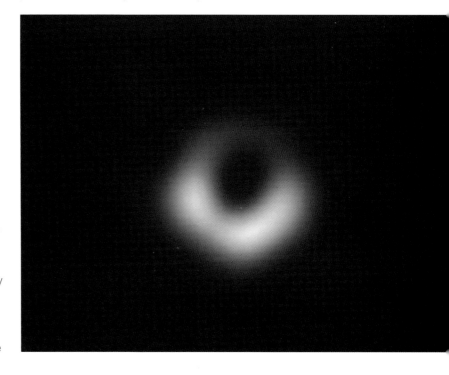

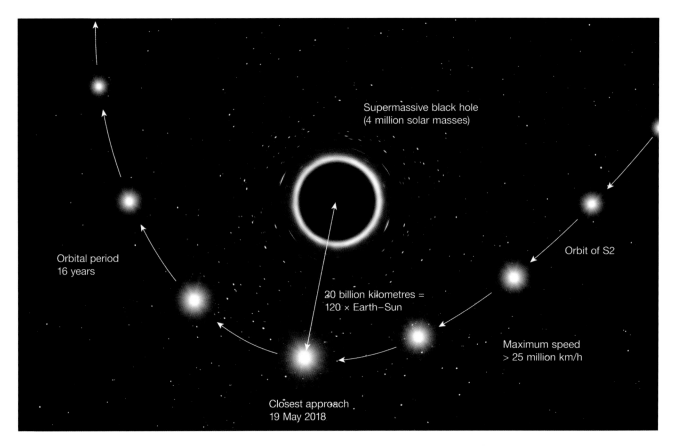

Supermassive black hole
(4 million solar masses)

Orbit of S2

Orbital period
16 years

20 billion kilometres =
120 × Earth–Sun

Maximum speed
> 25 million km/h

Closest approach
19 May 2018

ABOVE The path of the star S2 as it passes close to Sagittarius A*. As it gets closer, the cosmic heavyweight's strong gravitational field causes a disturbance in colour, turning from white to red, thanks to the effects of Einstein's General Theory of Relativity. *(ESO/M. Kommesser)*

BELOW The star S2 serves as a target in locating the position of the galactic centre, harbouring the supermassive black hole Sagittarius A*, which weighs in at 4 million solar masses (located by the orange cross). *(ESO/MPE/S. Gillessen et al.)*

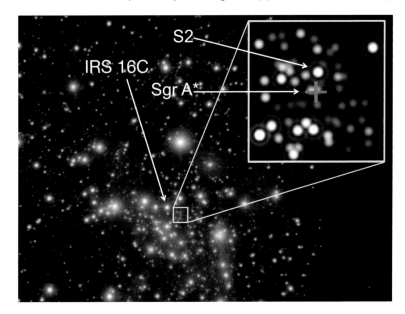

S2

IRS 16C

Sgr A*

of the Milky Way prevents us from seeing the region in the immediate vicinity of the black hole in visible light, but astronomers are currently (at the time of writing in the first half of 2019) trying to capture a radio-wavelength image of the black hole's event horizon, which is the point of no return from which not even light can escape the black hole, using a global network of radio observatories called the Event Horizon Telescope, which has already successfully imaged the supermassive black hole at the heart of the Messier 87 elliptical galaxy in the Virgo galaxy cluster, 54 million light years away.

A more indirect method of studying Sagittarius A* is to observe the motion of stars and gas clouds very close to the black hole. In particular, astronomers at the Max Planck Institute for Extraterrestrial Physics in Germany used the near-infrared CONICA instrument on the Very Large Telescope at the European Southern Observatory in Chile's Atacama Desert to see through the dust and observe a number of stars on their looping orbits around Sagittarius A*, including one very special star known only as S2.

S2 is special because astronomers have been observing it since 1995, tracking its

RIGHT Building a map using 1,339 large pulsating stars, each some 100,000 times brighter than our Sun, has revealed that the disc of our galaxy is warped – likely to be caused by the spinning of the Milky Way's inner stars. *(National Astronomical Observatories, Chinese Academy of Sciences)*

CENTRE An artist's impression of the G2 gas cloud passing close to the supermassive black hole at the centre of our galaxy. *(ESO/S.Gillessen/ MPE/Marc Schartmann)*

motion during its 16-year-long orbit around the black hole, during which it got as close as 17 light hours (the distance that light covers in 17 hours, which is about 18 billion kilometres (11 billion miles)) in 2002 and 2018. The star's velocity of up to 7,650 kilometres per second (4,753 miles per second), which is 2.5% of the speed of light, is only feasible if it is gripped by the gravitational field of a massive object that is millions of times more massive. Such an object can only be a black hole, thereby confirming the nature of Sagittarius A*. The observed radio emission is so-called synchrotron radiation, produced when high-speed electrons get caught up in the black hole's magnetic field, spiralling around the field lines and releasing radio-wavelength photons in the process. Similarly, a gas cloud named G2 passed close to Sagittarius A* in May 2014, and the Very Large Telescope was similarly used to follow its approach and observe how it interacted with the black hole's gravity. G2 swung around the black hole on what has been termed a 'supersquinched' orbit, in which it is hurled back from whence it came. Yet despite this perilous journey, G2 surprisingly remained intact against the fierce gravitational tides that exist around the black hole. For the gas cloud to have survived this passage rather than be torn apart, astronomers have suggested that G2 actually harbours a young star that is concealed inside it, and that the star's gravity is able to

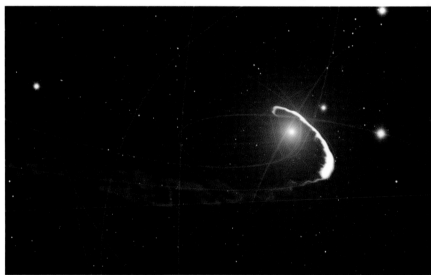

RIGHT By tracking the motions of the stars at the centre of our galaxy for more than 16 years, astronomers were able to determine the mass of its supermassive black hole. *(ESO/S. Gillessen)*

hold the gas cloud together. In 2018, three more similar objects were also discovered close to the black hole.

Some supermassive black holes seen in other galaxies are active, swallowing huge amounts of gas and spitting just as much back out into space. Around their event horizons form accretion discs of hot gas that swirl around the black hole's mouth. These discs can reach millions of degrees Celsius, and radiate brightly in optical light, ultraviolet and X-rays. As the disc of material swirls around and around, just a few million kilometres from the event horizon, magnetic fields that resided in the clouds of gas or stars torn apart to form the disc become tightly wound and ultimately funnel large amounts of the accretion disc material, now in its constituent parts of protons and electrons, away from the black hole in a tight beam accelerated to almost the speed of light. Indeed, such a beam can actually be seen coming from the black hole at the centre of Messier 87. The centres of galaxies containing active black holes such as this are known as active galactic nuclei, or AGN. They can come in several different flavours, the most powerful

of which are quasars, which are so active that we can see them across many billions of light years, all the way back to the distant era in which galaxies like our Milky Way were forming.

However, Sagittarius A* doesn't seem to be very active at all. In fact, today it is rather quiescent, at most emitting brief bursts of X-rays by frequently chomping down on rogue asteroids – NASA's orbiting Chandra X-ray Observatory has detected brief but frequent bursts of X-rays from the vicinity of the black hole, which has been interpreted as asteroids about 10km (6.21 miles) in diameter being gripped by the gravity of Sagittarius A*, and violently ripped apart and vaporised as they pass through the hot but diffuse gas around the black hole, like a meteor burning up in an atmosphere. The scientists who performed these particular observations with Chandra, led by Kastytis Zubov of the Center for Physical and Technological Sciences in Lithuania and formerly of the University of Leicester, propose that there is a cloud of trillions of asteroids swarming around Sagittarius A*, abducted from their parent stars when those stars passed too close to the black hole.

BELOW A composite image of the galaxy's core in infrared (red), near-infrared (yellow) and X-ray (blue and violet) wavebands reveals intense activity. *(Credit: X-ray: NASA/ CXC/UMass/D. Wang et al.; Optical: NASA/ESA/ STScI/D. Wang et al.; IR: NASA/JPL-Caltech/ SSC/S. Stolovy)*

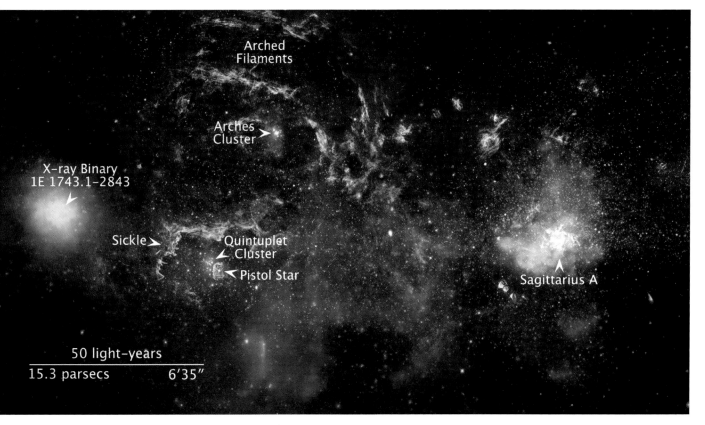

RIGHT Flares shooting from the black hole at the centre of the Milky Way have revealed that the behemoth has been vaporising some unlucky asteroids that have wandered into its path. *(NASA, CXC, MIT, F.K. Baganoff et al., E. Slawik)*

Those stars are mostly formed within the galactic centre, not too far away from the black hole. The galaxy's inner sanctum where these stars formed is called the Central Molecular Zone, or CMZ. Within the CMZ are several intense regions of star formation, such as the Arches Cluster. Located 100 light years from Sagittarius A*, the Arches Cluster is the most densely packed star cluster in the galaxy, but it wasn't discovered until 1995 because intervening gas and dust clouds block its visible light. Although infrared light can penetrate this gas and dust, hot, massive stars typical of those formed in the Arches Cluster don't shine brightly in infrared. Other star clusters within the CMZ include the Central Parsec Cluster and the Quintuplet Cluster. Stars such as the aforementioned S2 are believed to hail from these clusters.

In terms of numbers, the stellar populations of these CMZ clusters are dominated by low-mass stars. This is the same for every region of star formation in the known universe and is related to a property called the 'initial mass function', which describes the relative abundance of different masses of stars that form in stellar nurseries. The lower the mass of a kind of star, the more common that type of star is. Thus low-mass red dwarfs are most frequent, accounting for three-quarters of all the stars in the universe, while the least common are the massive stars that live short lives (roughly a few million years) and explode as supernovae. The Arches Cluster, for example,

RIGHT The Milky Way's centre is crammed with stars, as shown in this image snapped by the NASA/ESA Hubble Space Telescope. Using infrared wavebands, astronomers were able to peer through the dust that usually obscures our view. *(NASA, ESA and the Hubble Heritage Team (STScI/AURA). Acknowledgement: NASA, ESA, T. Do and A. Ghez (UCLA), and V. Bajaj (STScI))*

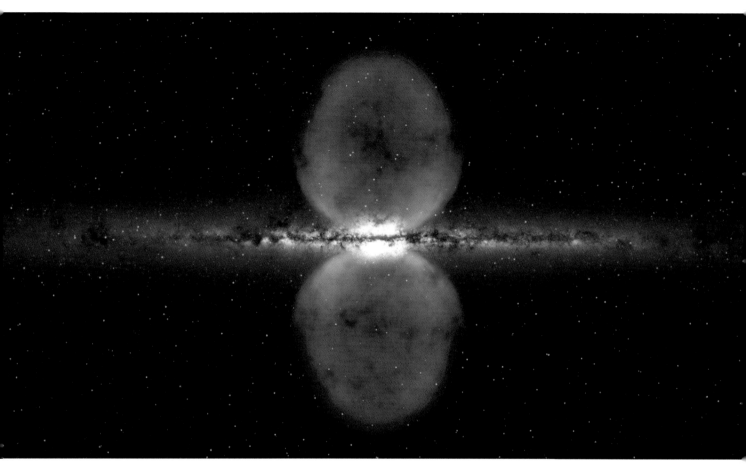

contains thousands of cool, low-mass stars, but only 135 hot, massive stars. Yet despite being outnumbered by more than ten to one, it's these stars that dominate the CMZ clusters, and which can influence the activity of Sagittarius A*.

Another Chandra X-ray study of the region around Sagittarius A*, this one run by Q. Daniel Wang of the University of Massachusetts at Armherst in the USA, has found that while the black hole does not have a dense accretion disc like that of an AGN, it is sipping on a small but steady stream of hot gas emitted on the stellar winds of hot massive stars like those found in the Arches Cluster. And it really is a meagre diet – only 1% of this stellar-emitted gas makes its way into the mouth of the black hole, while the rest escapes into space – the gas is too hot and energetic and diffuse for even the black hole's gravity to capture it all. This neatly explains why – apart from the regular destruction of passing asteroids – Sagittarius A* is faint in X-ray light, because the X-rays are emitted when the gas gets close to the event horizon and heats up,

but given that it is so sparse bar the occasional clump (some clumps of gas have been detected by the European Southern Observatory's Very Large Telescope orbiting close to the event horizon at about a third of the speed of light), there is therefore not much gas to heat up, and the low density reduces the friction between the atoms and molecules in the gas, reducing the heating further.

However, there is evidence that this was not always the case and that, once upon a time, the monster that is the black hole was awakened.

In 2010, astronomers Douglas Finkbeiner of the Harvard-Smithsonian Center for Astrophysics, and Tracy Slatyer and Meng Su of the Massachusetts Institute of Technology, found a new and mysterious structure in the galaxy that was not only extremely puzzling, but also extremely large, spanning a total of 50,000 light years above and below the galactic centre. The trio of astronomers had been making use of NASA's Fermi Space Telescope, which observes the cosmos in high-energy gamma rays, when they noticed two enormous plumes of gamma-ray-emitting material climbing high above and below the plane of the Milky Way, and centred on the location of the supermassive black hole. The best explanation for the existence of the plume is that, about three million years ago, Sagittarius A* experienced a dramatic and powerful outburst after it destroyed and consumed a massive object – perhaps a gas cloud thousands of times more massive than our Sun. For a short time, the centre of our galaxy would have been an AGN.

This kind of activity has important consequences for our galaxy as a whole, and also how Sagittarius A* formed in the first place. Bursts of radiation that blast out from around the black hole act to warm any interstellar molecular gas that the radiation may encounter. Molecular hydrogen gas clouds need to be extremely cold – less than -263°C (-441°F), which is about ten degrees above absolute zero – in order for gas molecules to clump together, fragment and gravitationally collapse. It is only when the gas collapses and the density rises that temperatures begin to grow, but by this stage gravity has taken over, attracting more gas and the core density and therefore temperature increases to the point that it becomes hot enough for nuclear fusion reactions to take place (as an example, the core temperature of the Sun is 15,000,000°C (27,000,000°F)). However, none of this can happen if radiation that is emitted from around the black hole heats the GMCs, and sometimes the radiation can even carry the molecular gas out of the galaxy entirely. Astronomers call this feedback, and to understand why, we need to understand how black holes form and grow.

Earlier in this chapter we introduced the notion of hierarchical formation, which is the idea that smaller chunks of galactic matter – dwarf galaxies, intergalactic gas clouds – come together at the nodes of the galactic web, surrounded by a halo of dark matter. However, scientists do not yet know if supermassive black holes formed in this way too. Did the Milky Way form first, and then the black hole inside of it, or does the formation of supermassive black holes predate the formation of galaxies? If it is the former, then it suggests that supermassive black holes form from the merger of lots of smaller black holes; if it is the latter, it could

be that supermassive black holes such as Sagittarius A* formed by the direct collapse of enormous gas clouds. Alternatively, it could be a combination of the two processes.

Within the dwarf galaxy NGC 4395, which is 14 million light years away, astronomers have detected a black hole with a mass about 10,000 times greater than the Sun – an intermediate-mass black hole – and the black hole in NGC 4395 is the largest intermediate-mass black hole yet detected. Intermediate-mass black holes are, of course, dwarfed by Sagittarius A*, but because NGC 4395 shows no evidence of being involved in any mergers in its history, and there is no evidence that the black hole has been active since it formed, the conclusion is that NGC 4395's black hole formed directly from the collapse of an immense cloud of gas, as opposed to smaller black holes, perhaps with hundreds or a thousand or so solar masses, colliding and merging.

Evidence for the merger scenario could

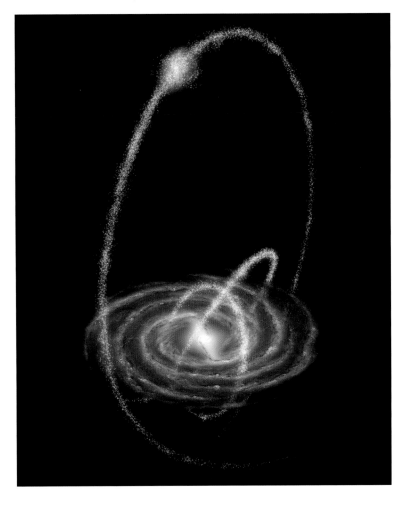

ABOVE The dwarf galaxy NGC 4395 is known to contain one of the smallest black holes, weighing in at only 300,000 solar masses. It possesses a halo and areas of significant brightness designated, and running from east to west, NGC 4401, 4400 and 4399. *(J. Schulman)*

BELOW The galaxy ESO 498-G5 contains a pseudobulge, which have the same appearance as conventional galaxy bulges but contain stars that orbit the centre in an orderly – rather than random – fashion. Our Milky Way is suspected to contain such a bulge. *(ESA/Hubble & NASA)*

come from detecting the gravitational waves emitted during black hole mergers. The first ever gravitational wave detection was made in 2015 by the Laser Interferometry Gravitational-wave Observatory (LIGO), which is based at two sites in the United States, at Hanford in Washington and at Livingston in Louisiana. Gravitational waves are ripples in the fabric of space-time that are produced by the interactions of extremely massive compact objects, such as neutron stars and black holes. The ripples wash through Earth, momentarily affecting the timing of laser beams that are bouncing back and forth between mirrors inside the LIGO detectors. However, all the black hole mergers that LIGO (now also joined by Europe's VIRGO detector in Italy) has detected so far are between stellar mass black holes, with masses several dozen times that of the Sun produced by the supernovae of the most massive stars. No mergers between black holes with hundreds or thousands of solar masses have yet been detected, although the higher the mass, the longer the wavelength of the gravitational waves, necessitating a future space-based gravitational wave detector, in the form of the Laser Interferometer Space Antenna (LISA), currently being developed by the European Space Agency.

However, with the evidence currently available, one could build a tentative picture of

the formation and growth of Sagittarius A*, first from the direct collapse of a gas cloud, and then either through a select few mergers, but mostly through the accretion of enormous amounts of gas, possibly going through a quasar stage very early in its life. Gas flows onto the black hole and feeds it, allowing it to grow by two orders of magnitude, from 10^4 solar masses to 10^6 solar masses. Eventually so much gas flows onto the black hole that it cannot consume it all, and radiation from the accretion disc starts to blow the infalling gas away, heating it so it can't form stars and on occasion ejecting it from the galaxy altogether. This is why it's called feedback – the flow of material that grows the black hole is ultimately its own downfall, acting to shut off that infall. Sporadic feeding episodes may then occur in the billions of years afterwards, as evidenced by the Fermi Bubbles, but overall Sagittarius A* and the Milky Way galaxy around it have settled into a relative state of quiescence, declining star formation and overall maturity.

Future evolution

For a time – a long time – secular evolution will dominate in the Milky Way galaxy, as it will now in most galaxies, since the expansion of the universe has increased the distance between many gravitationally unbound galaxies, reducing the regularity of major galaxy mergers.

ABOVE One of the Laser Interferometer Gravitational-wave Observatory (LIGO) detector sites at Livingston, Louisiana. The other resides at Hanford in eastern Washington and together they measure gravitational waves. *(Caltech/MIT/LIGO Lab)*

However, there are still plenty of large galaxies that are gravitationally bound, and one day the Milky Way's quiescent state is going to be rudely disrupted.

We'll explore this further in Chapter seven, but it is worth briefly alluding to here as an epilogue to the story of the formation and evolution of the Milky Way so far.

The Milky Way exists as part of the Local Group of galaxies – a collection of three large spirals, the Milky Way, the Andromeda Galaxy and the Triangulum Galaxy, and over 50 dwarf galaxies spanning a volume of space about 10 million light years across. The three spirals are engaged in a deadly gravitational dance. The Milky Way and the Andromeda Galaxy are approaching one another at 110 kilometres per second (68 miles per second), and in about 4.5 billion years, they will crash into one another. When that happens, both galaxies will be transformed forever. The collision will send stars spinning into all kinds of orbits, while the vast GMCs will collide, instigating a new round of frantic star formation. In the end, we will be left with a gigantic elliptical galaxy, nicknamed 'Milkomeda'.

Chapter Three

Our galactic neighbourhood

The Milky Way is not alone in space. It has a small family of galaxies, known collectively as the Local Group. The main members, besides the Milky Way, are the famous Andromeda Galaxy (also known as Messier 31, or just M31) and the Triangulum Galaxy (Messier 33, or M33). Both are spirals, but each is very different to the other, and different to the Milky Way, providing an excellent opportunity to compare and contrast and better understand the characteristics of our galactic home.

OPPOSITE Andromeda in infrared, captured by the Wide-field Infrared Survey Explorer, or WISE for short. *(NASA/JPL-Caltech/UCLA)*

57

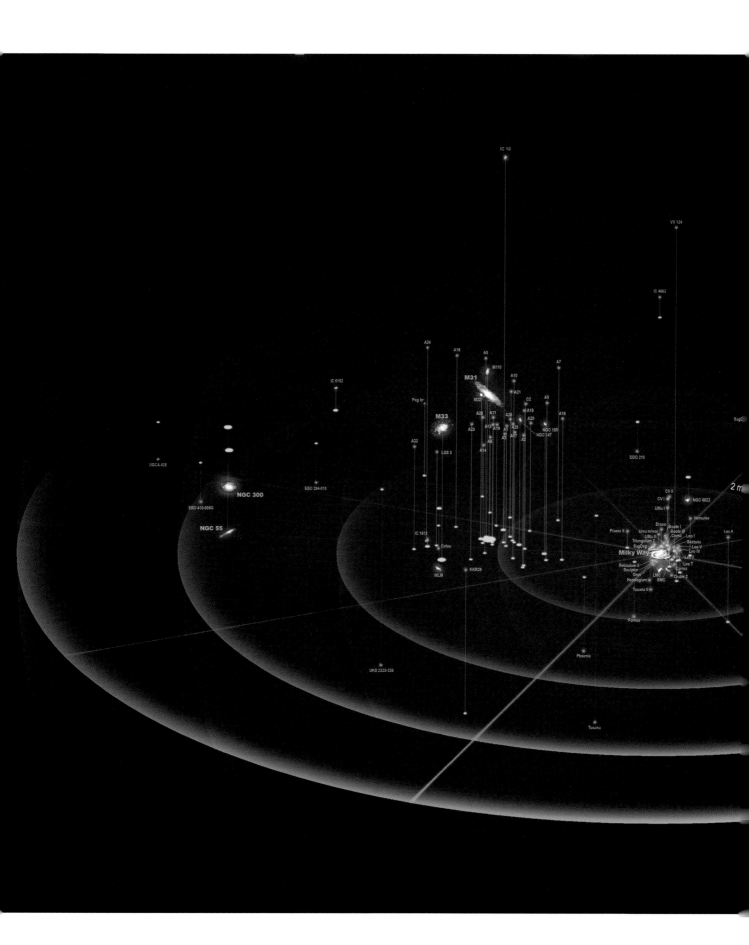

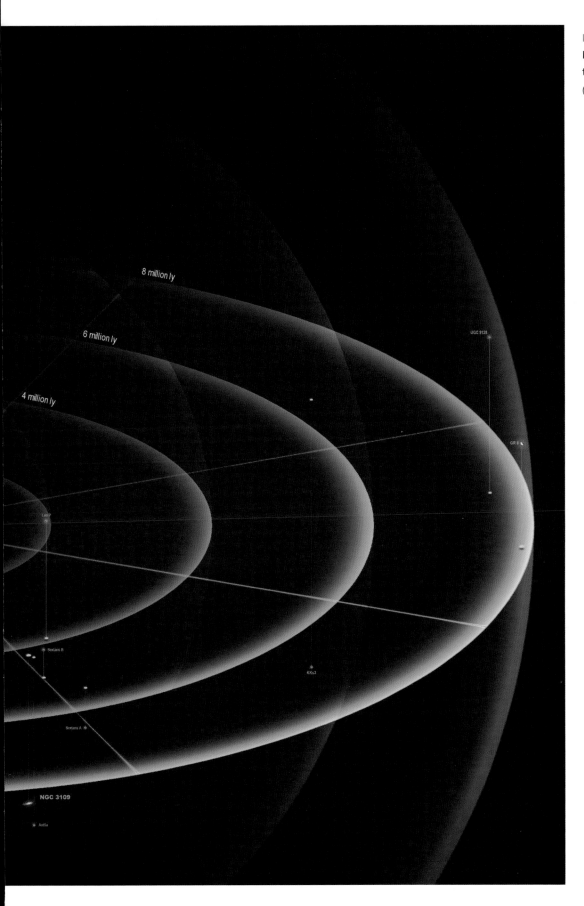

LEFT A map of the
Local Group, in which
the Milky Way resides.
(Antonio Ciccolella)

8 million ly

6 million ly

4 million ly

UGC 9128

GR 8

Leo P

Sextans B

KKs3

Sextans A

NGC 3109

Antlia

Alongside these large spirals are over 50 dwarf galaxies; we don't know how many for sure because models of hierarchical galaxy formation predict that there should be many more, but if they exist then they so far remain undetected. Of the ones we do know about, the most notable are the Large and Small Magellanic Clouds, as well as Messier 32 and Messier 110 around the Andromeda Galaxy.

Most of the Local Group's galaxies are clustered into two groups: the Milky Way and its retinue of satellite galaxies, and Messier 31, Messier 33 and their associated dwarfs. The two groups are gravitationally connected, and are gradually moving towards one another. The exceptions are a small group featuring the dwarf galaxies NGC 3109, Sextans A and B, and the Antlia dwarf galaxy, which exist on the edge of the Local Group.

The Andromeda Galaxy

No large spiral galaxy beyond the Milky Way has been studied as well as the Andromeda Galaxy. Its sheer size, at 200,000 light years across it is twice as big as the Milky Way, and proximity (at a distance of 2.5 million light years it is the nearest large galaxy to us)

provides the perfect opportunity to explore another galaxy from a distance. An enormous 1.5 billion pixel image showing just a portion of M31's core and spiral structure, taken by the Hubble Space Telescope and released in 2015, shows just how detailed we can go in our observations of the Andromeda Galaxy – the image features 100 million individual stars, and shows dust lanes and gas clouds. Some of those stars are Cepheid variables, which Edwin Hubble used to measure the distance to the Andromeda Galaxy in 1924 using the historic 2.5m (8ft 2in) Hooker Telescope on Mount Wilson in California, discovering that it is so far away that it – and the other puzzling 'spiral nebulae' – were in fact galaxies in their own right, 'island universes' beyond our Milky Way. That single realisation completely changed our understanding of our place in the cosmos, showing the universe to be far bigger than we could have ever imagined.

From Earth, we see the Andromeda Galaxy at a highly inclined angle of 77 degrees, meaning that we view it more from an edge-on perspective than face-on. This can cause some observational difficulties as astronomers try to grasp the true shape of its disc. However, observations at infrared wavelengths, which

BELOW A bird's-eye view of a section of the Andromeda Galaxy in the sharpest image ever taken of the spiral by the Hubble Space Telescope. *(NASA, ESA, J. Dalcanton, B. F, Williams and L. C. Johnson (University of Washington), the PHAT team, and R. Gendler)*

allow astronomers to see through the dust in the disc, reveal some surprises.

First is that M31 is a barred spiral, much like our Milky Way is. This was only realised in 2006, following infrared observations as part of research led by Racheal Beaton, now at Princeton University, using the Two Micron All Sky Survey (2MASS). The observations showed that M31's inner bulge is relatively spherical,

ABOVE The European Space Agency (ESA)'s space telescopes – Herschel and XMM-Newton – in an alternative light: infrared and X-rays. *(ESA/ Herschel/PACS/SPIRE/J. Fritz, U. Gent/XMM-Newton/EPIC/W. Pietsch, MPE)*

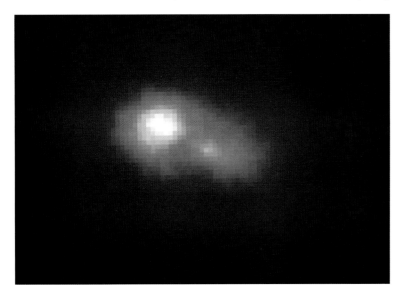

but with increasing distance from the centre of the Andromeda Galaxy, the bulge becomes boxier. This happens because we are seeing the vertical profile of the bar, close to edge-on as a result of the high inclination of M31's disc.

However, like the Milky Way, the processes that went into building M31's bulge and bar remain a little murky. A study in 2018 led by Roberto Philip Saglia, of the Max Planck Institute for Extraterrestrial Physics in Germany, used the VIRUS-W spectrograph on the 2.7m (8ft 10in) telescope at the McDonald Observatory in Texas to spectroscopically explore the characteristics of the stars in M31's bulge. They found that the stars in both the bulge and the bar are over 10 billion years old, so one might think that they are Population II stars, but their average metallicities match the Sun's abundance of heavy elements. The scientists' conclusion is that M31's central regions formed in two stages. First, a classical bulge formed through a process of mergers and gravitational collapse, followed by the formation of a 'proto-disc' around the bulge that over time grew unstable and buckled, developing the bar that evolved into the boxy bulge.

At the heart of that bulge is an unusual double nucleus, composed of two bright regions separated by five light years. The clumps are named P1 and P2, in order of brightness. P2 is actually located at the very centre of the Andromeda Galaxy and is the galaxy's supermassive black hole, which contains about 60 million solar masses, making it a real heavyweight compared to Sagittarius A*. Observations performed by the Hubble Space Telescope reveal a bright disc encircling P2, in which the far brighter P1 is embedded. This is thought to be a brilliant star cluster filled with massive and luminous stars. The cluster has likely been gravitationally captured – perhaps it is the core remnant of a dwarf galaxy that was cannibalised by M31 long ago – since the gravitational and thermal environment would be too extreme for stars to form so close to the black hole.

Beyond the bar and the bulge is the Andromeda Galaxy's spiral disc, which has an

LEFT The Andromeda Galaxy features a double nucleus. *(NASA/ESA)*

interesting story to tell. It's not easy to see in visible light, but when the Andromeda Galaxy is observed at other wavelengths, such as in ultraviolet light by NASA's Galaxy Evolution Explorer (GALEX) spacecraft, and especially in infrared images taken by the likes of NASA's Spitzer Space Telescope and the European Space Agency (ESA)'s Herschel Space Observatory, M31's disc seems to consist of a series of concentric rings that appear slightly off centre. They are understood to have been created by a cosmic bullet: the nearby and aforementioned satellite galaxy M32, which is suspected of having ploughed through M31's disc about 900 million years ago. As it punctured the disc and careened through, it generated ripples in M31's disc like throwing a stone into a pond, producing the concentric rings. The entry point was initially suspected to be just to the left of the core of the galaxy as seen in the images, and this was thought to be the reason for the lopsidedness. However, when simulating this collision in a computer, Avi Loeb and Marion Dierickx, both of the Harvard-Smithsonian Center for Astrophysics, and Laura Blecha of the University of Maryland, found that impact occurred near the edge of M31's disc, and that the concentric ripples produced by M32 are tightly wound spiral arms. These spiral arms were therefore formed by the shockwaves of M32's passage through the galactic disc. The impact also would have led to M31's S-shaped

ABOVE A mosaic of our closest spiral in ultraviolet. *(NASA/ Swift/Stefan Immler (GSFC) and Erin Grand (UMCP))*

BELOW Andromeda in infrared, captured by the Spitzer Space Telescope. *(NASA)*

warp, in contrast to the secular processes that have produced the Milky Way galaxy's warp.

Messier 32 is a highly compact dwarf elliptical galaxy, with a diameter of just 6,500 light years but containing 3 billion solar masses of material, and 2.5 million solar masses of that material is contributed by a supermassive black hole – astonishingly it is more than half the mass of Sagittarius A*, despite the diminutive nature of M32. Furthermore, its stars are rather similar in character to the stars in the bulges of spiral galaxies. One hypothesis was that M32 is the core of a long-lost spiral galaxy that lost its spiral arms due to an earlier interaction, quite possibly a glancing blow, with the Andromeda Galaxy about two billion years ago. However, the computer simulations by Loeb's group dispute this, implying that no such glancing blow took place, and that M32 was already a dwarf elliptical when it plunged through M31's disc.

The other main satellite galaxy of M32, namely M110, is a more typical dwarf galaxy, of the spheroidal variety. It has a diameter of 16,000 light years and a total mass ten billion times greater than our Sun. It is estimated that most of its star formation ceased half a billion years ago, but in the past 20 million years, something – likely gravitational tides emanating from the Andromeda Galaxy – has stirred up the molecular hydrogen gas in M110 once more, sparking a new round of star formation. Several of the hot, massive stars formed in this recent burst of star formation have since exploded as supernovae – five supernovae have been recorded in M110. Other dwarf spheroidal galaxies orbiting the Andromeda Galaxy include NGC 147 and NGC 185.

Despite being a dwarf, M110 has its own retinue of globular clusters, numbering at least a dozen. However, this number – and even the 150 globular clusters known to orbit our Milky Way galaxy – are dwarfed by the population of globular clusters orbiting the Andromeda Galaxy. Over 460 have been identified so far, the brightest of which is known as G1 (which stands for 'Globular 1') and which contains around 17 million solar masses, far greater in mass than any globular clusters around the Milky Way. There is also evidence for an intermediate-mass black hole inside G1, which points to the cluster probably being the remnant core of a dwarf galaxy rather than being a bona fide globular cluster.

The Triangulum Galaxy

Whereas the Andromeda Galaxy is highly inclined to us, the Triangulum Galaxy (M33) is aligned at a more forgiving angle, giving astronomers a clearer view of its spiral structure. Although different methods have given different results for M33's inclination, most range between 40 and 55 degrees.

The Triangulum Galaxy is about 60,000 light years across, so it is the smallest by far

RIGHT A view of the Andromeda Galaxy's globular cluster G1, also known as Mayall II, which contains at least 300,000 ancient stars. *(Michael Rich, Kenneth Mighell and James D. Neill (Columbia University), and Wendy Freedman (Carnegie Observatories), and NASA/ESA)*

SPOTTING THE TRIANGULUM GALAXY IN THE NIGHT SKY

The Triangulum Galaxy is 2.9 million light years away, containing about 40 billion stars, and on a very dark, clear night in autumn, you might just be able to see it as a vague, fuzzy patch in the sky, shining at magnitude 5.7 at the edge of naked-eye visibility. If you do manage to see it without an optical aid, then congratulations – you've just seen the most distant object visible to the naked eye.

of the three large spiral galaxies in the Local Group. If we look at M33 visually, it does appear substantially different in structure compared to the Milky Way galaxy and M31. It has a bulge, but it is rather diminutive and, unusually for a spiral galaxy, no supermassive black hole has been found in the centre of M33. What does exist in its centre are several enormous and highly luminous star clusters, plus the most powerful and persistent X-ray source in the entire Local Group. It's called M33 X-8, and it lies smack bang at the centre of the Triangulum Galaxy. It's thought to be a black hole, but not a supermassive black hole as exists in the Milky Way and the Andromeda Galaxy – M33 doesn't appear to have a supermassive black hole, or even an intermediate-mass black hole for that matter. Instead, M33 X-8 is what astronomers call an X-ray binary system, which features a compact object – which is sometimes a neutron star but which in this case is a 14 solar mass black hole – accreting gas from an expansive, bloated red giant companion star that is in orbit around it. A red giant is one of the later evolutionary phases of a star like our Sun, and its diffuse outer layers are susceptible to being torn away by the gravity of the black hole. The red giant's material swirls around the black hole in an accretion disc, becoming so hot and dense as to produce X-rays. Indeed, the X-ray emission is so powerful that it accounts for 70% of the Triangulum Galaxy's entire X-ray output.

It's purely by coincidence that M33 X-8 finds itself at the centre of the Triangulum Galaxy. In lieu of a supermassive black hole, the inner region is populated by several massive star clusters, one of which M33 X-8 is likely a member of. While studies have differed on the exact age of the stars in the relatively tiny nucleus, the observations do seem to indicate that there have been several main population bursts. Some stars are several billion years old, while another population is younger than half a billion years old, and the youngest stars

BELOW This is a cloud of dust and stars situated in the Triangulum Galaxy, also known as Messier 33. It's 2.9 million light years away from Earth and glows with the energy of hundreds of bright, young stars. *(ESA/Hubble and NASA)*

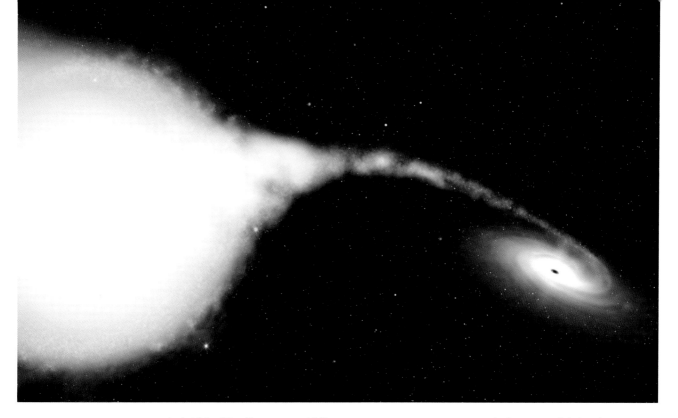

ABOVE An artist's impression of an X-ray binary, where material is being gravitationally pulled off a supergiant star by a black hole. *(ESA)*

are just 40 to 50 million years old (the age can be determined via studies of the star's colour, which is a proxy for temperature, and the star's luminosity, and then comparing these with models of stellar evolution, since stars grow brighter and hotter as they age – we'll discuss this more in Chapter four).

It is suspected that the nucleus of M33 formed via gas flowing inwards from the spiral disc. Initially, the gas formed a small disc in the nucleus, before expanding outwards above and below the plane of the nuclear disc to form the flattened spheroid shape, similar to the shape of a classic galactic bulge, but smaller.

Meanwhile, the Triangulum Galaxy's spiral disc has its own story to tell. One glance at the galaxy is enough to tell us that it appears differently to the Andromeda Galaxy. There are two main spiral arms, but there are many messy-looking spurs emanating from those arms, leading to an appearance that is described as 'flocculent', as opposed to the more streamlined appearance of so-called 'grand design' spirals such as the Whirlpool Galaxy (Messier 51) in the constellation of Canes Venatici, or Bode's Galaxy in Ursa Major (Messier 81). This more fluffy, patchy appearance is thought to be the result of 'self-propagating' star formation, whereby star formation in one region produces stars with powerful radiation winds, which eventually go

supernova, producing powerful shockwaves that can stir up neighbouring molecular gas clouds and cause them to form more stars, which produce their own winds and shocks, which stir up more gas, and so on and so forth. It is not clear how much influence density waves have in forming the arms of flocculent spirals such as M33, but models have been produced that combine the two processes to result in something akin to M33's appearance.

Studies of the abundance of heavy elements within the stars and gas in the Triangulum Galaxy's spiral disc has also taught us that the disc formed from the inside-out. The inner part of the disc contains stars that are, on average, about 10 billion years old, while the outer disc contains younger stars, and in general they follow a pattern of decreasing age with increasing distance from the centre of the galaxy.

Another peculiarity in the disc is a warp, not in the stellar disc, but in the disc of neutral hydrogen gas that pervades the galaxy and extends out from the centre of M33 by a greater distance than the disc of stars. The two discs are misaligned with each other by about 30 degrees, possibly the result of a close passage of the Andromeda Galaxy in the past. Neutral hydrogen radiates at the radio wavelength of 21cm, meaning that neutral hydrogen is best observed with radio telescopes. In 2016, researchers led by Olivia Keenan of Cardiff University found something unexpected when analysing data from the Arecibo Galaxy Environment Survey, or AGES for short, which is a set of observations performed by the giant 300m (984ft) dish of the Arecibo radio telescope in the jungles of Puerto Rico to map neutral hydrogen in the Local Group. Around the Triangulum Galaxy are numerous hydrogen gas clouds that don't contain any stars, so they aren't dwarf galaxies, and actually one of these clouds is so big that its diameter is equivalent, or perhaps even greater, than the stellar disc of M33. The cloud, called AGESM33-31, has a mass about 12 million times greater than our Sun, a diameter of at least 60,000 light years, and is shaped like a ring doughnut. Its origin remains unknown.

How many globular clusters the Triangulum Galaxy possesses is also currently uncertain. Around 50 have been positively identified, but unlike the Milky Way or the Andromeda Galaxy, there doesn't seem to be as clear a delineation in the age, colour or luminosity of globular clusters found in the galaxy's small bulge and halo and those large star clusters found in the stellar disc. Instead, unless the location of any given cluster is clearly seen to be in the halo (from our perspective it can be difficult to determine the location of a cluster in the galaxy when seen on the two-dimensional night sky), then the main way of differentiating between them is looking at their motions – true globular clusters are on random orbits around the galaxy, whereas objects embedded within the galaxy's stellar disc orbit with the plane of that disc. Some of the unusually dense star clusters within the disc are referred to as 'young populous clusters', although 'young' may be a misnomer, since their age is intermediate, ranging from a few billion years to 7 billion years old. On the other hand, if the Milky Way and the Andromeda Galaxy are anything to go by, then true globular clusters are 12 to 13 billion years old. These massive, dense, spheroidal clusters in the stellar disc are not seen in either the Milky Way or Andromeda (although as we shall see shortly, they are found in the Magellanic Clouds), but perhaps one way to explain them is the presence of immense star formation within the Triangulum Galaxy. The biggest star-forming nebula in M33 is NGC 604, which is the second most massive and luminous star-forming region in the entire Local Group, with a diameter of 1,500 light years. If it were located at the same distance as the Orion Nebula, which is the closest large region of star formation to the solar system at a distance of 1,344 light years, then NGC 604 would appear 6,300 times brighter and outshine the planet Venus in our sky. It contains over 200 massive stars that will one day each explode as a supernova, and the entire cluster of stars within NGC 604 is estimated to be about 3 million years old. Only such intense star formation as in regions like NGC 604 could produce the kinds of young populous clusters as seen in the Triangulum Galaxy.

The Magellanic Clouds

The Milky Way's closest galactic neighbours of note are the two Magellanic Clouds, named after the Portuguese explorer Ferdinand Magellan, whose expedition into the Southern Hemisphere while circumnavigating the world 'discovered' them in the early 16th century (of course, they were known to the indigenous people of the Southern Hemisphere long before then). To the naked eye they appear as fuzzy clouds, hence their names, but in reality they are a pair of dwarf galaxies located between 162,000 and 200,000 light years away. Only the Sagittarius Dwarf Spheroidal Galaxy, at a distance of 52,000 light years, is closer to the Milky Way. However, that is so close that the Milky Way's gravity is pulling the Sagittarius

BELOW Our impressive Milky Way, the Carina Nebula glowing intensely at its centre. The Large Magellanic Cloud is visible at the top right, while the Small Magellanic Cloud is at the right of the image. *(ESO/H. Stockebrand)*

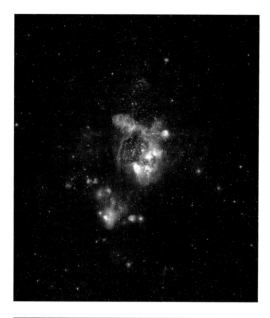

Dwarf Spheroidal apart, leaving a stream of stars behind it. It's even been suggested that the disc instability that produced the Milky Way's spiral arms resulted from gravitational interactions with the Sagittarius Dwarf Spheroidal Galaxy as it plunged close to, or perhaps even through, the Milky Way's disc.

But let's get back to the Magellanic Clouds. The Small Magellanic Cloud (SMC) is the more distant of the two, is 7,000 light years across and possesses a total mass of 7 billion solar masses, most of which is dark matter. Meanwhile, the Large Magellanic Cloud (LMC) spans 14,000 light years across, with a mass of the order of tens of billions of solar masses. It is host to 60 globular clusters, 400 planetary nebulae (nebulae form when Sun-like stars die) and 700 open star clusters (which are young star clusters like those that form in the discs of galaxies). Despite the LMC's small size, it is actually a distorted dwarf spiral galaxy, complete with a central bar and at least one distorted spiral arm. The distortions are the product of gravitational tides resulting from interactions with the SMC and the Milky Way galaxy, and the bar is also warped, or bowed, with the two ends closer to the Milky Way than the middle of the bar.

For now, however, it is the interactions between the two Magellanic Clouds that wield the greatest influence on their appearance and structure. For billions of years the two dwarf galaxies have been gravitationally attacking one

ABOVE The remnants of a long-dead star named **DEM L316A** and located some 160,000 light years away in the Large Magellanic Cloud. *(ESA/Hubble & NASA, Y. Chu)*

ABOVE RIGHT One of the biggest globular clusters in the Large Magellanic Cloud, NGC 1783. *(ESA/ Hubble & NASA Acknowledgement: Judy Schmidt)*

BELOW The Small Magellanic Cloud (SMC). *(ESA/Hubble and Digitized Sky Survey 2)*

BELOW RIGHT The Milky Way's gravity tugs on its satellite, the Large Magellanic Cloud, causing gas clouds to collapse and young stars to be born. *(NASA, ESA, Acknowledgement: Josh Lake)*

ABOVE A small section of the Large Magellanic Cloud, boasting a collection of baby stars, most weighing in at less than the mass of the Sun. The young stellar cluster is known as **LH63**. *(NASA, ESA and D. Gouliermis (University of Heidelberg), Acknowledgement: Luca Limatola)*

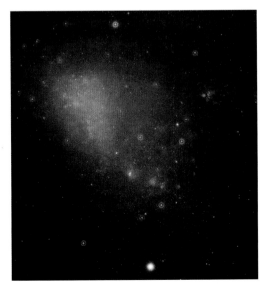

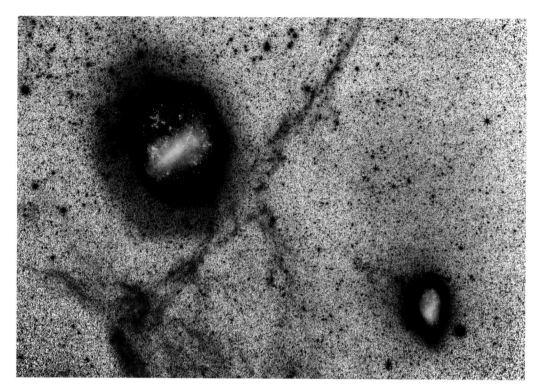

another, and even bumping into each other from time to time. The repeated interactions produce tidal forces that stretch and compress the gas in each galaxy, instigating the formation of yet more star clusters. Gravity has caused the outer regions of the SMC to become stretched such that they now point towards and away from the LMC, while the part of the LMC that is closest to the SMC has become warped. The most damning evidence for their interactions is how

the 'Wing' – the south-eastern part of the SMC – has literally taken off and become separated from the main body of the galaxy. It's now moving away in the same direction as the LMC, following a collision between the two galaxies in the past two billion years.

Then there's the Magellanic Stream, which is a river of gas and stars flowing away from the its smaller galactic sibling and connecting to the LMC, whose gravity is pulling the stream away from the SMC, like pulling on a string rolled up in a ball of wool, and depleting the SMC of future star-forming molecular hydrogen gas. The Magellanic Stream rings the Milky Way galaxy, spanning 100 degrees of the sky, and there is so much dense molecular gas in the Stream that it is actually forming stars, outside of any galaxy. There's even an additional small galaxy – an ultra-faint dwarf that has very few stars and is dominated by dark matter – named Hydrus I, which was discovered by a team led by Dougal Mackey of the Australian National University and which lies within the Magellanic Stream, possibly as a satellite of one or both of the Magellanic Clouds.

The repeated interactions between the two over cosmic history, resulting in the distorted structures and starbursts and the Magellanic Stream, make for fascinating study, especially as they are so close, allowing astronomers to

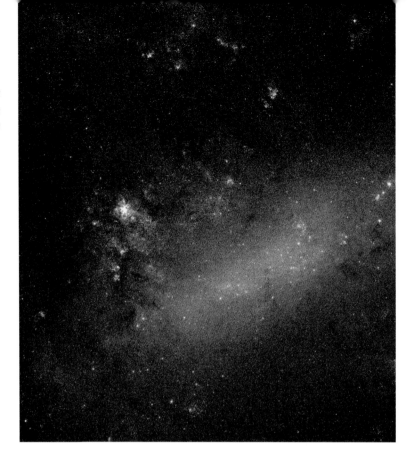

RIGHT The Large Magellanic Cloud snapped by the European Southern Observatory (ESO)'s 1m (39in) Schmidt telescope. *(ESO)*

get a high-resolution view of what's happening to them. However, in about 2 billion years' time, we'll get an even closer view.

For a long time there was a great debate about whether the two Magellanic Clouds are in orbit around the Milky Way, or whether they are just 'passing through', with a velocity high enough to take them out of the clutches of the Milky Way's gravity. Ultimately, however, it turns out that it doesn't really matter which is true, because either way their fate is sealed. New findings have determined that the LMC contains about twice as much dark matter as had previously been thought, which changes its own gravitational field, and how it gravitationally interacts with the Milky Way. When this new data was input into the powerful EAGLE (Evolution and Assembly of GaLaxies and their Environments) supercomputer, astronomers at the University of Durham discovered that in about 2 billion years' time, the LMC will collide with, and merge into, the Milky Way. Its stars will be integrated into our galaxy, while its gas clouds will collide with GMCs in our galaxy, spurring a burst of fresh star formation, a precursor to the big event a few billion years later when the Andromeda Galaxy will crash into

the Milky Way. It is thought that the SMC will also follow suit some undetermined time later.

It wouldn't be the first merger that the Magellanic Clouds will have been involved in. Evidence that there was once a third Magellanic Cloud, that was consumed by the LMC between three and five billion years ago, has come to light. While measuring the motions of the stars in the LMC, Benjamin Armstrong of the International Centre for Radio Astronomy Research in Australia

BELOW The Magellanic Stream. The Large Magellanic Cloud and Small Magellanic Cloud form the head of this long ribbon of gas, which stretches nearly halfway around the Milky Way. *(NASA)*

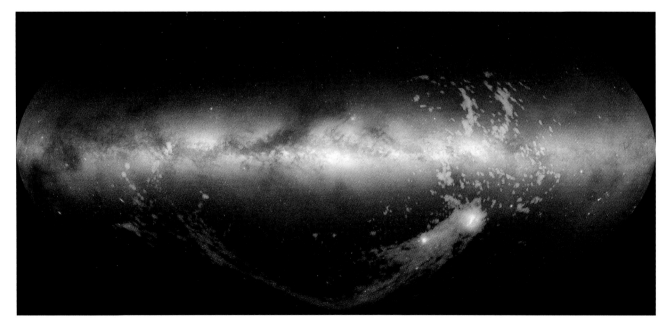

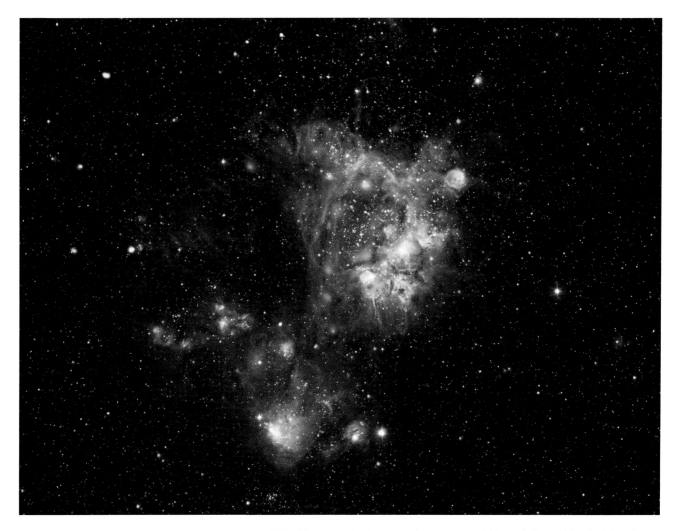

ABOVE Colourful view of the star-forming region LHA 120-N44 in the Large Magellanic Cloud. The centre reveals a very rich region of gas, dust and young stars known as NGC 1929. *(Optical: ESO, X-ray: NASA/CXC/U. Mich./S. Oey, IR: NASA/JPL)*

OPPOSITE The incredible star-forming region R136 in the large open cluster NGC 2070. *(NASA, ESA, F. Paresce (INAF-IASF, Bologna, Italy), R. O'Connell (University of Virginia, Charlottesville) and the Wide-Field Camera 3 Science Oversight Committee)*

found that while most of its stars orbit clockwise around the galaxy's centre, some are orbiting counter-clockwise. This can sometimes occur when a smaller galaxy merges with a larger galaxy at an angle opposite to the larger galaxy's direction of rotation; the stars from the smaller galaxy that become integrated into the larger galaxy end up orbiting in the opposite direction. The counter-clockwise stars are therefore a smoking gun that a merger took place in the past. Armstrong even suspects that the merger can explain the LMC's 'age-gap' problem, where its star clusters are either very old, or fairly young, with nothing in between, as though the LMC had shut down its star formation for billions of years.

A merger would have helped kick-start the star-formation process again after a lengthy hiatus, while the subsequent interactions with the SMC then maintained that star formation through to the present day.

And nowhere in the entire Local Group is forming stars more intensely than the LMC's Tarantula Nebula, also known as 30 Doradus. It's an immense region of star formation, up to 1,800 light years across, very similar in size to NGC 604 in the Triangulum Galaxy. The best way to compare them, however, is to again imagine how bright they would be if placed at the distance of the Orion Nebula. Whereas we've seen that NGC 604 would be brighter than Venus in our skies, the Tarantula Nebula would be bright enough to cast shadows!

One of the reasons that the Tarantula Nebula is so active is that it is located on the leading edge of the LMC, where the dwarf galaxy collides with intergalactic gas that fills the space

OUR GALACTIC NEIGHBOURHOOD

R136A1 – THE MASSIVELY UNSTABLE STAR

R136a1 has a mass 315 times greater than the Sun, and a luminosity that is an incredible 8.7 million times brighter than the Sun. Compare this to the most luminous stars in the Milky Way galaxy, namely the Pistol Star in the central Quintuplet Cluster, LBV 1806-20 near the galactic centre and Eta Carinae, which is 7,000 light years away. These three stars have total luminosities that are 1.6 million, 2 million and 1 million times greater than the Sun's luminosity, respectively.

R136a1 actually started out even more massive than the Tarantula Nebula, estimated to be 325 solar masses, but because it is so massive it is highly unstable and pressure from radiation produced by nuclear reactions in its core is enough to lift huge amounts of gas from its outer layers, blowing it away in a powerful stellar wind and depleting its overall mass by ten solar masses so far, and by more in the future. Even so, losing all this mass is not going to prevent R136a1 from meeting its ultimate fate in a cataclysmic supernova explosion.

between galaxies. In a way, for the galaxies, moving through the intergalactic medium is a bit like wading through very hot treacle, and the pressure on the LMC's leading edge as it ploughs through this intergalactic gas causes the compression and collapse of gas clouds within the LMC, sparking star formation.

Several generations of stars have already been born, lived and died in the Tarantula Nebula even as it continues to form new stars. Such intense star-forming regions have a propensity to produce more massive stars that are highly luminous (and therefore have powerful radiation winds) and soon explode as supernovae after a few million years. At the centre of the Tarantula Nebula is the massive star cluster NGC 2070, which has a total mass of 450,000 solar masses, and which contains the most massive star ever found: R136a1. As

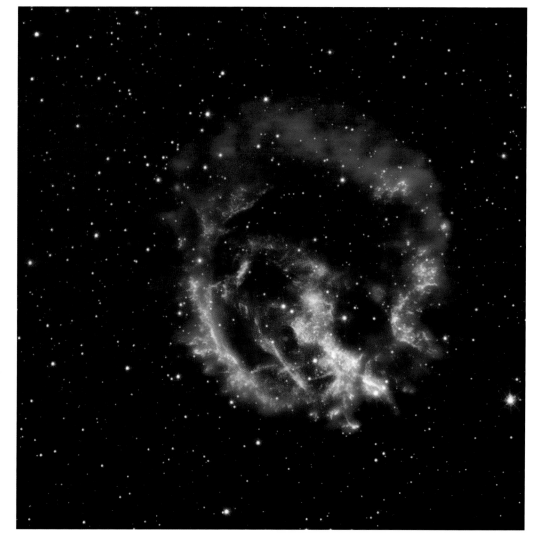

RIGHT The supernova remnant 1E 0102.2-7219 in the Small Magellanic Cloud, made visible in optical and X-ray wavebands. *(ESO/ NASA, ESA and the Hubble Heritage Team (STScI/AURA)/F. Vogt et al.)*

for NGC 2070, when all is said and done, it will likely turn into one of the young populous clusters seen in the Triangulum Galaxy.

We know that the LMC is a hotbed for supernovae. Numerous supernovae remnants have been identified, particularly N49, which is an expanding, X-ray emitting cloud of stellar debris produced by a supernova that exploded about 5,000 years ago. More recent was supernova (SN) 1987A, which occurred when a blue supergiant star called Sanduleak -69 202 was seen to explode on 23 February 1987. It was so bright that it could clearly be seen with the naked eye (from the Southern Hemisphere only, of course, where the LMC is visible). The supernova was an incredibly important event in the annals of modern astronomy – it was the first (and so far only) time that modern telescopes have been able to study a supernova so closely (most are seen to explode in galaxies millions of light years away), watching the event unfold and even detect a burst of neutrinos, which are tiny, almost massless particles, from the supernova, which provided scientists with information about the nuclear physics taking place within the star as it exploded. Indeed, they confirmed theories that of all the energy released by the supernova, which amounts to more than 4×10^{46} joules, 99 per cent of that energy was in the form of neutrinos. Since SN 1987A's explosion, the Hubble Space Telescope has tracked the development of the expanding debris, which has begun to collide with circumstellar material (gas and dust that had previously been ejected by the doomed star) causing it to light up in a series of concentric rings. Today, the supernova's shockwave is expanding through space at a velocity of 3,600 kilometres per second (2,237 miles per second).

Our place in the universe

The Local Group is the Milky Way's home, but it is only one small part of a much larger story. Around 54 million light years away is another complex of galaxies, known as the Virgo Cluster, which is much, much larger than the Local Group. In fact, the Local Group is gravitationally bound to the Virgo Cluster, an outlying satellite in a similar fashion to the

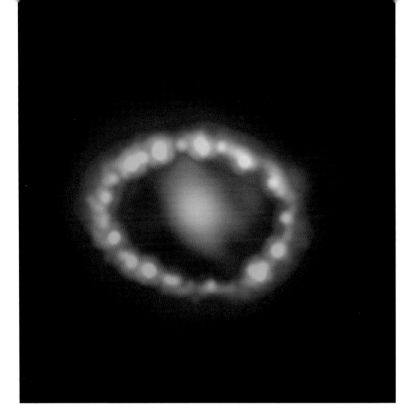

ABOVE The remnant of Supernova 1987A – an expanding shockwave can be seen colliding with a ring of material around the exploding star.
(ALMA (ESO/NAOJ/NRAO)/A. Angelich. Visible light image: the NASA/ESA Hubble Space Telescope. X-Ray image: The NASA Chandra X-Ray Observatory)

BELOW It might look like a lonely galaxy, but Wolf-Lundmark-Melotte, or WLM for short, is considered to be part of our Local Group.
(ESO Acknowledgement: VST/Omegacam Local Group Survey)

dwarfs that orbit the Milky Way galaxy from a distance. Together, the Virgo Cluster and its satellite groups, including the Local Group, form the Virgo Supercluster.

The Virgo Cluster itself is home to at least 1,300 galaxies, some of which are massive elliptical galaxies, such as Messier 87, which is located at the very heart of the cluster. There are spiral galaxies too, but they tend not to fare

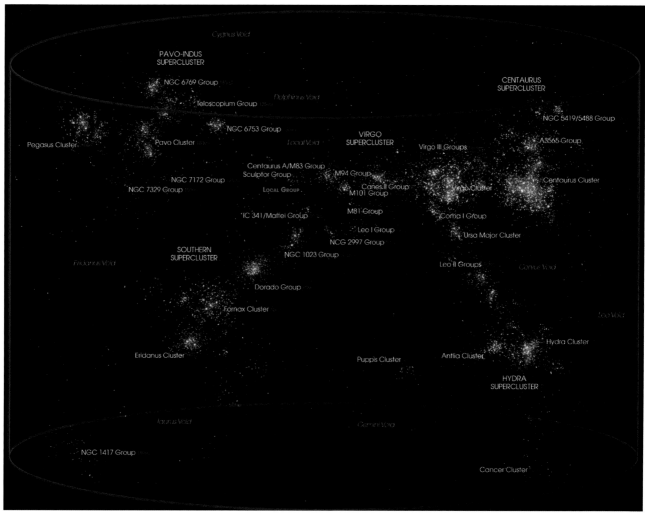

well in the extreme environment of the cluster. To understand why, we first need to understand how galaxy clusters form. In Chapter 2 we introduced the idea of the cosmic web of matter that fills the universe. Galaxies form along the filaments of this web. The Local Group formed at a small node between several filaments, but this is linked to an even larger node where lots of filaments intersect to form lots of galaxies in a cluster. As more and more gas and galaxies, attracted by gravity, flow along the filaments and enter the cluster, the cluster grows over billions of years. As spiral galaxies moving along these filaments ultimately fall into the Virgo Cluster – or any large cluster for that matter – they encounter the 'intra-cluster medium'; a fog of extremely hot hydrogen gas, even hotter and more dense than the intergalactic gas being encountered by the LMC. As the spiral galaxies plough through this gas on their way down the gravitational well to the core of the cluster, the hot gas scours them, stripping cold gas out of them, or heating that gas so it can no longer form stars. Known as ram-pressure stripping, this turns once fertile gas-rich galaxies into the dusty discs of lenticular galaxies that are described as 'red and dead' (red being the colour of older, smaller, longer-lived stars, whereas blue signifies hot but short-lived stars signifying recent star formation). Some of these lenticular galaxies will collide, forming elliptical galaxies like those at the heart of the Virgo Cluster.

Yet the Virgo Supercluster is not the be all and end all when it comes to galaxy conurbations. Instead it is just a part of an even larger structure, called the Laniakea Supercluster. It combines not only the Virgo Supercluster, but three other superclusters and their sub-clusters – up to 500 known galaxy groups – in an incredibly large assembly that spans 520 million light years in diameter and contains approximately 100,000 galaxies. Although Laniakea, which means 'immense heaven' in Hawaiian, is not gravitationally bound, meaning that the expansion of the universe will ultimately separate its component superclusters, all its member clusters share similar motions through space and are interacting gravitationally to some extent. Other superclusters on the same scale at Laniakea are also known to exist in the universe. For the

point of view of this book, they serve to remind us of the Milky Way's importance in the grand scheme of things: the Milky Way is an average spiral galaxy, one of hundreds of billions of galaxies in the known universe. The Milky Way resides in a humdrum group of galaxies, which orbits a much larger galaxy cluster, which in turn is but a subset of an even bigger structure that takes light half a billion years to cross. The Milky Way is not at the centre of the universe, or even the centre of a cluster. It's not even the biggest galaxy in the Local Group. It's the ultimate depiction of Copernicanism, that the Milky Way's location and existence is nothing special in the grand scheme of things. Except that, to us, it is something special: it's our home galaxy, and in our next chapter we'll explore the Sun's local neighbourhood within the Milky Way, and find out whether we reside in a special place in our galaxy that enables life to survive on Earth.

BELOW The Abell 901/902 supercluster, which is located roughly 2 billion light years away, kick-started the knowledge into the existence of superclusters in 1958. (ESO)

Sun

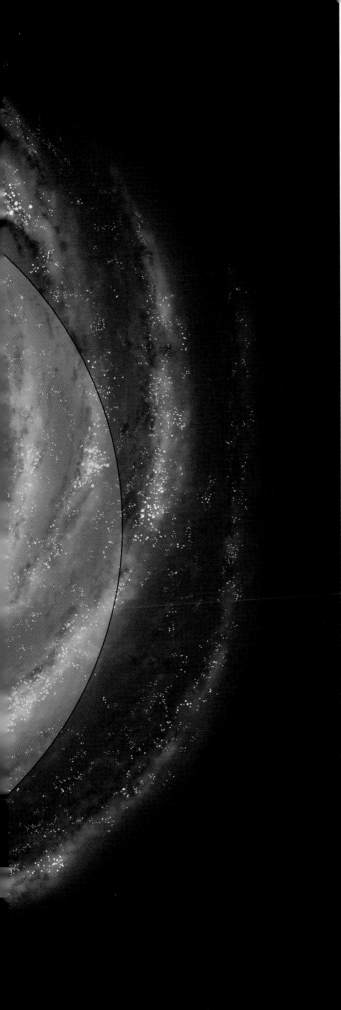

Chapter Four

Our place in the Milky Way

The answer to exactly how many stars reside in the Milky Way may vary depending on which astronomer you speak to, with estimates ranging between 100 billion and 400 billion, although many observers settle on about 200 billion stars. Either way, that's a lot of stars, and our Sun is but one such object among all these, currently living off the beaten track in a nondescript arm of the galaxy.

OPPOSITE The Galactic Habitable Zone of the Milky Way – a section of the galaxy where life is theoretically able to exist. *(NASA/Caltech)*

Yet there could be something special about the Sun's location in the galaxy. We are here after all; perhaps the chance that life can develop on a planet is all about location, location, location. It's a concept known as the Galactic Habitable Zone, and it can tell us a lot about the varying properties across the breadth of the Milky Way's disc.

The Galactic Habitable Zone

The idea is that, for a planet to form life, it requires certain ingredients: heavy elements such as carbon, oxygen, nitrogen, phosphorous, calcium, iron and so forth. These ingredients are produced inside stars through various nuclear reactions. For example, inside the Sun, carbon is produced at the end of a lengthy chain of reactions. First the Sun has to produce helium by what is called the proton-proton chain, whereby two hydrogen nuclei (protons) are fused together in the Sun's core to form helium-2, which then decays into deuterium (an isotope of hydrogen), that fuses with another helium nuclei to produce helium-3. From this stage, there are several paths that the helium-3 can take to form a full helium nucleus, helium-4, but then this sets the stage for the 'Triple-Alpha Process' (helium-4 nuclei are referred to as alpha particles), in which two helium-4 nuclei first fuse to create a beryllium-8

nuclei, and then this fuses with another helium-4 nucleus to produce carbon-12. Heavier elements are built through similar but increasingly complex reactions. A Sun-like star will run out of nuclear fuel and expand into a red giant before gently releasing most of its mass to form a planetary nebula. These expanding nebulae effectively represent the star shedding its skin, and elements formed inside the star, including carbon, oxygen and nitrogen, are also released. More massive stars have higher core temperatures and pressures and can maintain fusion reactions to form even heavier elements, all the way up to iron in the star's core. To then fuse iron nuclei into heavier elements requires more energy than the fusion reaction produces, so it is at this stage that fusion reactions within the star cease and the star explodes as a supernova, spewing its guts, including the heavy elements it has produced, all across space. Only in the incredible heat and energy of the supernova explosion are elements heavier than iron created – in fact, supernovae produce most of the rest of the elements in the periodic table (the exception being elements with atomic numbers 95 or greater, which have so far only been produced in the laboratory).

So elements that make planets and water, and cells and bones, all come from stars. As Carl Sagan used to say, we are made of starstuff. Therefore, the Galactic Habitable Zone – that is, the part of the Milky Way that has the best conditions for life – will out of necessity need to have enough of these heavy elements, so it needs to be in a region where there have been enough generations of stars that have lived and died and recycled their elements into the next generation to produce sufficient elements for planets and life to form.

In general, within the Milky Way's disc, the abundance of heavy elements decreases with radius from the galactic centre. Get too far out, on the edge of the spiral arms, and there may not be enough elements to build many planets or facilitate complex life.

Going the other way, towards the galactic centre, and the density of stars increases. In particular, the density of star formation is greater, with all those intense star-forming zones within the inner disc and galactic bar, and the centre of the galaxy. More star formation means

BELOW G1.9+0.3 is the remains of a supernova that is the most recent of all time, exploding within the lifetime of the Earth. It is located roughly 28,000 light years from us. *(X-ray (NASA/CXC/ NCSU/K. Borkowski et al.); Optical (DSS))*

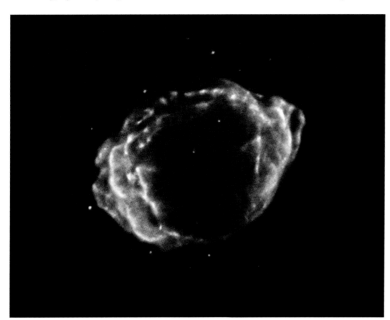

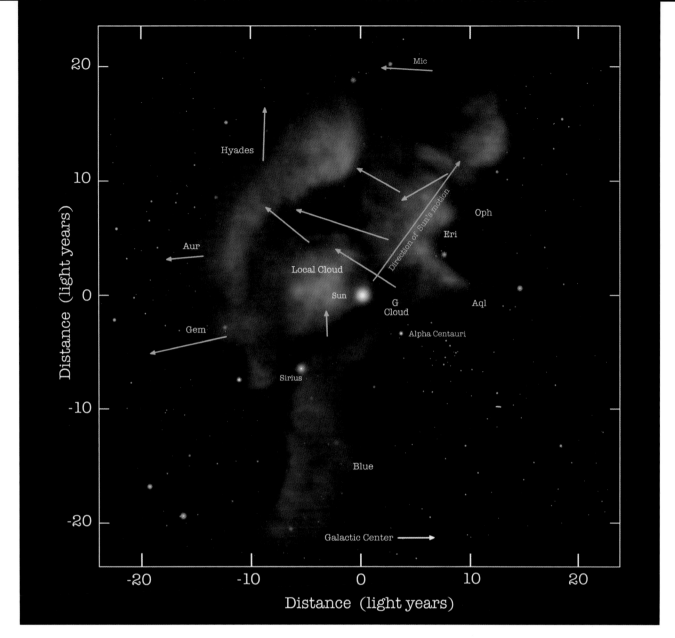

Distance (light years)

more massive stars that inevitably explode as supernovae. The radiation from a supernova can affect planets within a volume several hundred light years across – for instance, recent work by Adrian Melott of the University of Kansas has suggested that a supernova that exploded 2.6 million years ago at a distance of 300 light years from Earth resulted in a mass extinction that killed off 36 per cent of marine genera (i.e. the types of life that live in the ocean).

So, supernovae can be deadly, and life would not really be able to get a foothold if there were stars exploding on too regular a basis. Based on the galactic abundance of the radioactive element aluminium-26, which is produced in supernovae, the Milky Way experiences, on average, one supernova every 50 years, and while some do occur in

the disc, most of these are concentrated in the galactic centre (for example, the youngest known supernovae remnant, named G1.9+0.3, exploded near the centre of the galaxy sometime between 1890 and 1908, but its light was hidden by intervening interstellar gas and dust, and the remnant was only subsequently detected in X-ray and radio wavelengths).

The supernova rate would seem to rule out wide swathes of the centre of the Milky Way, while the dearth of heavy elements makes the outer regions of the galaxy's disc less appealing for life. That leaves a broad ring around the galaxy, with an inner boundary 13,000 light years from the galactic centre, and an outer boundary of about 32,000 light years (compared to the radius of the Milky Way, which is approximately 50,000 light years). This

ABOVE Our journey through space is taking us through a gathering of very low density clouds of interstellar gas and dust. *(NASA/Goddard/ Adler/U. Chicago/ Wesleyan)*

is the purported Galactic Habitable Zone, and, happily, the Sun and solar system sits in the middle of this ring, at a distance of 25,900 light years from the galactic centre.

The Galactic Habitable Zone doesn't mean that life can't form in other parts of the galaxy, it only suggests that life is less likely to form and survive in those areas. But let's zoom in on our little neighbourhood in the galaxy, and see what kind of a place we live in.

The Local Bubble

Between the stars is the interstellar medium (ISM), which is a kind of thin, gaseous gruel filled with atoms and molecules, particles of dust and hard-hitting cosmic rays, which are high-energy particles unleashed by the most magnetically energetic processes in the universe. The ISM is clumpy, with hotter, sparser voids in between cooler, denser patches. The Sun is currently passing through one of these

emptier voids, known as the Local Bubble, which has an atomic density of just 0.05 atoms per cubic centimetre, around a thousand times less dense than the average density of the ISM in the Milky Way's spiral disc.

The Local Bubble is not spherical. As best as we can tell, it is shaped a bit like a peanut shell, and it is thought that the Bubble is a cavity carved by the expanding shockwave of a supernova, or supernovae, that exploded on the scene in the past 20 million years. Our Sun and the solar system was not even inhabiting the Local Bubble at the time; the Sun is moving around the galactic centre at a velocity of 230 kilometres per second (143 miles per second), taking about 230 million years to complete one orbit, and on its journey as it travels around the galaxy, bobbing up and down in the plane of the thin disc, it has wandered into the slower-moving Local Bubble.

Being the product of potentially multiple supernovae, the remaining gas inside the Local

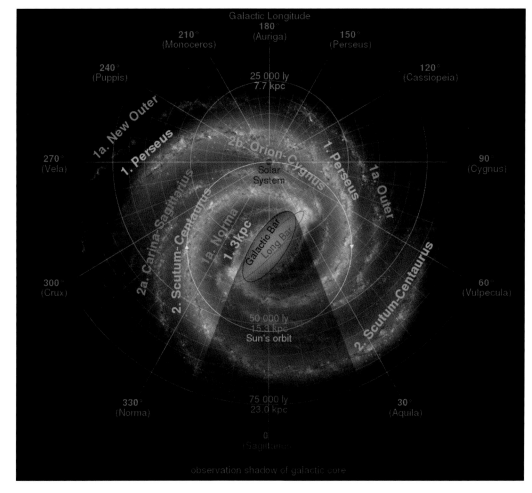

RIGHT The path of our Sun around the Milky Way – our solar system takes between 225 to 250 million years to complete one orbit, travelling at an average speed of 828,000 kilometres per hour (514,000 miles per hour). *(NASA/JPL-Caltech/R. Hurt)*

Bubble is extremely hot, millions of degrees Celsius, although this is mitigated somewhat by the extremely low density of material. However, even inside the Local Bubble, the ISM is patchy, and these denser clumps have much lower temperatures of several thousand degrees Celsius. Our solar system is skirting along the perimeter of one of these slightly denser patches that has been nicknamed the Local Fluff (more properly known as the Local Interstellar Cloud). It's about 30 light years in diameter, and is currently being sampled by the Voyager 1 and 2 spacecraft, which were launched by NASA all the way back in 1977. They were designed to fly past and observe the outer planets of the solar system – Jupiter, Saturn, Uranus and Neptune – and their moons. Voyager 1 only encountered Jupiter and Saturn, with mission planners back on Earth opting for it to take a trajectory that took it close to Saturn's large moon Titan but also out of the plane of the solar system, while Voyager 2 sped onwards to also explore Uranus and Neptune.

After passing Neptune in 1989, Voyager 2, along with Voyager 1, remained operational. Although some of their instruments, such as their cameras, have been switched off to preserve power, they retained their ability to detect particles and electromagnetic fields, and monitored how the gusts of the Sun's solar wind began to slow in the face of the ISM. The point at which the solar wind stops is known as the heliopause, which is the boundary to a magnetic bubble blown by the Sun called the heliosphere. Beyond the heliopause lies the ISM. Voyager 1 crossed the heliopause and entered interstellar space on 25 August 2012. Voyager 2, lagging behind a little bit and heading on a different trajectory to Voyager 1, followed suit on 5 November 2018. Both are now interstellar explorers, but space is so vast that it will take them thousands of years to get anywhere. For example, in about 40,000 years, Voyager 1 will pass within 1.6 light years of the star Gliese 445,

RIGHT Beyond the orbit of ice giant Neptune, the solar wind spewed by our Sun interacts with the interstellar medium to create the inner heliosheath. It's bounded by a termination shock on the inside and the heliopause on the outside. *(NASA/IBEX/Adler Planetarium)*

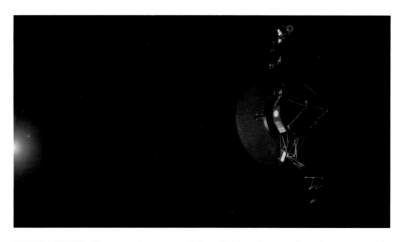

ABOVE NASA's Voyager 1 spacecraft headed into interstellar space in 2013. Its twin left the solar system in 2018. *(NASA/JPL-Caltech)*

BELOW The Interstellar Boundary Explorer (IBEX)'s construction of the first-ever sky map of the interactions at the edge of the solar system. Our Sun's influence diminishes and interacts with the medium in our solar system – the interstellar boundary region protects us from the dangerous cosmic radiation from our galaxy. *(NASA/Goddard Space Flight Center Scientific Visualization Studio/Tom Bridgman)*

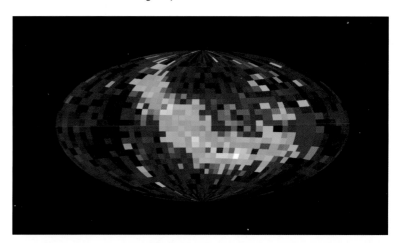

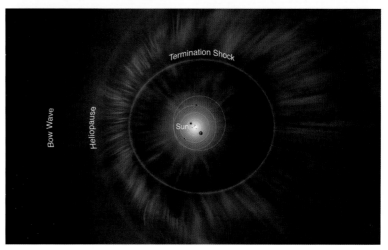

RIGHT Just to the right of the dome of La Silla Observatory, Alpha (right) and Beta Centauri (left) are visible. *(Y. Beletsky (LCO)/ESO)*

RIGHT At 4.37 light years away, Alpha Centauri is the closest star system and planetary system to us. Proxima Centauri is visible as the fainter red dwarf in this image. *(ESO/DSS 2)*

which is currently 17.6 light years away but in 40,000 years' time will be only 1.8 light years away, such is the star's motion through space. Of course, by then both Voyager spacecraft will be inoperative – their power reserves will last only until the mid-2020s.

BELOW LEFT Sirius is a binary star system and the brightest star in the night sky. It rests 8.6 light years away. *(NASA, ESA and G. Bacon (STScI))*

BELOW Hubble Space Telescope image of Sirius A and smaller, white dwarf companion Sirius B. *(NASA, ESA, H. Bond (STScI) and M. Barstow (University of Leicester))*

As for our Sun and solar system, as we move along the edge of the Local Fluff, we are heading towards another, larger patch of the ISM called the G-Cloud. The nearest star system to us – specifically the Proxima and Alpha Centauri grouping that are 4.2 and 4.3 light years away respectively – are inside the G-Cloud, which shows how close it is to us. Meanwhile, the brightest star in the night sky, namely the Dog Star Sirius, is 8.6 light years behind us, outside of the Local Fluff. In about 20,000 years' time the Sun will pass from the wisps of the Local Fluff and into the G-Cloud.

The length of time that it will take the Sun, and the Voyager spacecraft, to travel any significant distance through space gives us an idea as to how big the Local Fluff, the G-Cloud and the Local Bubble itself are relative to us, but in the grand scheme of things, compared to the size of the Milky Way galaxy, they are very small.

The Warren

The Local Bubble is bumping into another large interstellar cavity called Loop I, which contains the impressive Scorpius-Centaurus Association of hot, massive stars, some of which have already gone supernova, creating the North Polar Spur that emanates from Loop I and extends high above the plane of the galaxy, about 490 light years from the Local Bubble. Other supernova-driven bubbles have also been identified nearby, and between them

are 'chimneys' – interconnected tunnels of denser gas that weave through the ISM like a rabbit warren, having formed when expanding bubbles collide with colder clumps in the ISM, the shockwave on the expanding edge of the supernova-induced bubbles compressing the clumps into thin shells, which then encounter other bubbles, breaking up to form the chimneys. It is thought possible that the entirety of the Milky Way could be filled by this network of bubbles and chimneys.

All of these features are carved by supernovae explosions. We can see similar phenomena in other galaxies, such as the Magellanic Clouds, where astronomers bear witness to 'super-bubbles', which can be clearly seen as huge cavities filled with hot,

ABOVE At a distance of 8.6 light years, Sirius – also known as Alpha Canis Majoris – is one of the closest stars to Earth. Here, it is visible on the bottom of the image, forming the Winter Triangle with red supergiant Betelgeuse (top right) and binary system Procyon (top left). *(Akira Fujii)*

LEFT The Scorpius-Centaurus Association is the nearest OB association to Earth. The largest stars in the region contain decent amounts of carbon monoxide, offering new insight into how giant planets form and evolve around young stars. *(NRAO/AUI/NSF; D. Berry/SkyWorks)*

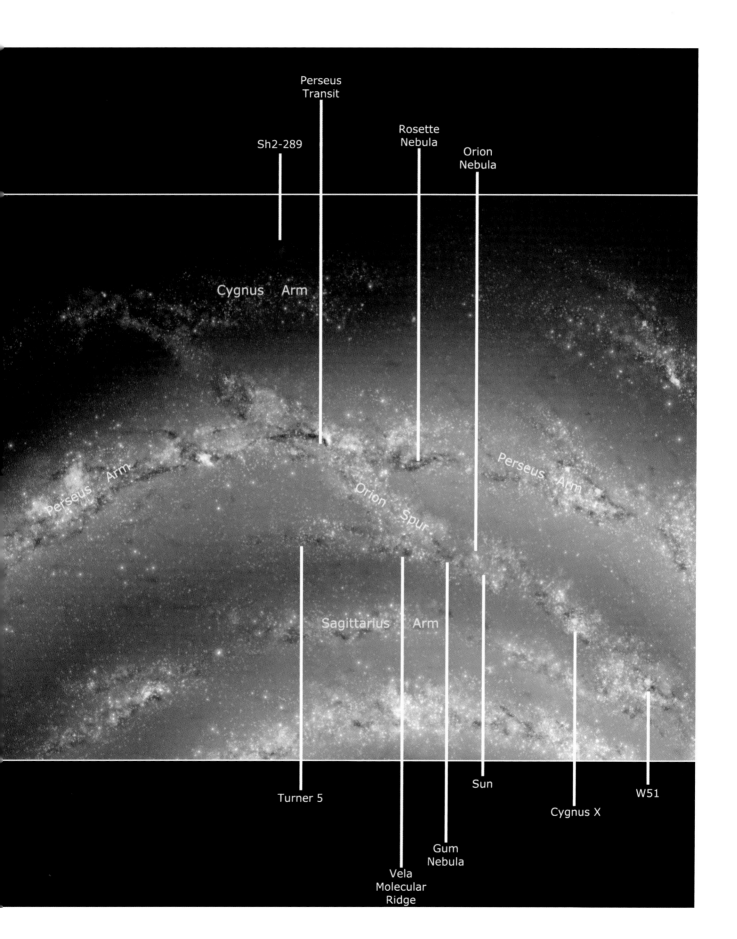

Perseus
Transit

Sh2-289

Rosette
Nebula

Orion
Nebula

Cygnus   Arm

Perseus   Arm

Perseus   Arm

Orion   Spur

Sagittarius   Arm

Turner 5

Sun

W51

Cygnus X

Gum
Nebula

Vela
Molecular
Ridge

OPPOSITE A close-up of the Orion Spur, in which the Solar System resides. *(R. Hurt)*

ionised gas expanding into denser nebulosity. However, many of these super-bubbles are found in giant, active star-forming regions, where massive stars aplenty are forming, and the bubbles themselves are set against a background of molecular gas. For the Local Bubble, its neighbour Loop I and others in the local neighbourhood, things aren't quite as dramatic, but astronomers are trying to work out where the supernovae that blew these bubbles were located. The attention has come to focus on a giant ring of star formation that surrounds the bubbles, called Gould's Belt, named after the 19th century astronomer, Benjamin Gould.

Gould's Belt spans about 2,000 light years in diameter, and the Sun is just 326 light years from its geometric centre, so that the ring literally rings around us in the night sky. It's characterised by giant molecular clouds including the famous Orion Nebula, and groupings of hot, massive stars known as OB Associations ('O' and 'B' referring to the two hottest categories of stars) such

ABOVE The Dumbbell Nebula (Messier 27) resides in the Orion Arm. *(Göran Nilsson & The Liverpool Telescope)*

LEFT The Beehive Cluster (Messier 44) is a member of the Orion Arm. *(Miguel Garcia)*

as the aforementioned Scorpius-Centaurus Association. The 30 million-year-old Belt is tilted by 17 degrees to the plane of the Milky Way, and is suspected to originate from the impact of a giant 10 million-solar-mass cloud of intergalactic gas and dark matter, moving at high speed, that plunged through the disc of the Milky Way. As it did so, it set off ripples, like when you drop a stone into a pond, and these ripples compressed molecular gas clouds around the rim of the impact point, instigating a circle of star formation. Scientists estimate that such collisions occur in the Milky Way every 300 million years or so, meaning that the galaxy should be dotted with other Gould Belts, although their landmark bright stars and star-forming nebulae are likely long since gone. However, like our presence in the Local Bubble,

ABOVE Messier 7 is easily spotted with the naked eye in the direction of the tail of the constellation of Scorpius. *(ESO)*

RIGHT The Orion Nebula's little sibling, Messier 43 is often referred to as De Mairan's Nebula. It forms part of the Orion molecular cloud complex. *(ESA/Hubble & NASA)*

ABOVE The spectacular Messier 47 is a young open cluster imaged by the 2.2m (7ft 2in) telescope at ESO's La Silla Observatory in Chile. *(ESO)*

ABOVE RIGHT An artist's impression of a hot Jupiter discovered in the star cluster Messier 67, which rests in our section of the galaxy – the Orion Arm. *(ESO/L. Calçada)*

the Sun's existence inside Gould's Belt is purely coincidental as we just happen to be passing through – the Sun was not on the scene when the intergalactic cloud struck.

Gould's Belt is one of the most significant features in the Orion Spur, which is the modest little spiral arm that the Sun finds itself a member of. We say 'little', but it is still 10,000 light years long and 3,500 light years wide. It is considered to probably be a segment of spiral structure branching off the Perseus spiral arm, which is one of the Milky Way's two main spiral arms. Along with the Sun, the Orion Arm contains some of the best and most well-known celestial objects visible in the night sky: the

RIGHT The Little Dumbbell Nebula (Messier 76) is one of many Messier objects that reside in the Orion Arm. *(Daniel Nobre/Liverpool Telescope)*

Orion Nebula, the Dumbbell planetary nebula, the Ring planetary nebula, and young star clusters including the Pleiades and the Beehive.

Birthplace of the Sun

Despite all the star formation ongoing in Gould's Belt and in the Orion Arm as a whole, the Sun and its solar system did not form in its current position in the Milky Way. As we have alluded to, the Sun is merely passing through the Local Bubble, and as we learned in Chapter 2, the spiral arms of so-called 'grand design' spiral galaxies are produced when stars and gas bunch up in density waves as they orbit their galaxy. The Sun has orbited the Milky Way 20.4 times since it was formed within a molecular gas cloud, one perhaps similar to the Orion Nebula. In that time, the other stars that formed with it have all drifted apart. We

can see this slowly happening in open star clusters, which are young clusters of stars, a few hundred million years old, that inhabit the galactic thin disc. The aforementioned Pleiades and the Beehive, among many others, are classic examples. The main stars of the Plough (or Big Dipper as it is known in the USA) are described as a co-moving group – they were once part of a cluster, but have since spread out while sharing the same motion.

Since stars in an open cluster all formed at the same time, it is therefore relatively simple to measure the cluster's age, using a fantastic tool called the Hertzsprung–Russell (HR) diagram. Developed in the early 20th century by astronomers Ejnar Hertzsprung and Henry Norris Russell, it plots the luminosity of a star against its spectral type, or colour (which is a proxy for temperature). Most stars are in what is called the Main Sequence, which describes stars

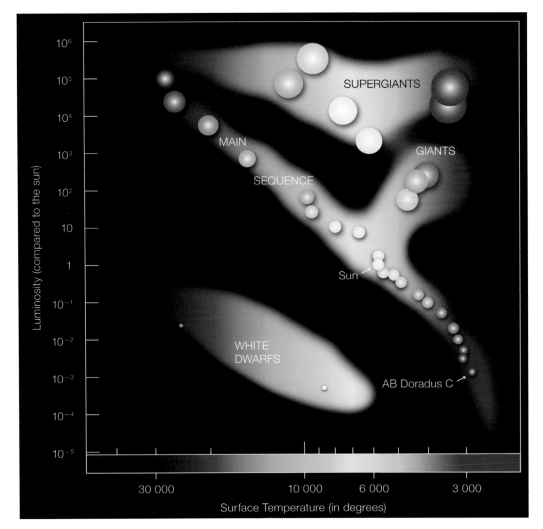

that still fuse hydrogen in the nuclear reactions inside their cores to generate energy that makes a star shine. In the HR diagram, the Main Sequence is a band running from bottom right (low luminosity, low temperature) to top left (high luminosity, high temperature). Once a star runs out of hydrogen – and the more massive stars run out of it faster – then the star moves off the Main Sequence and onto either the so-called 'Giant Branch', wherein they expand to become red giants, or if they have masses greater than eight times the mass of the Sun, they then instead move onto the 'Supergiant Branch', where they eventually explode as supernovae.

By placing the stars in a cluster onto an HR diagram, astronomers can see how far up the Main Sequence a star has evolved, identify which stars have turned off the Main Sequence and onto either the Giant or Supergiant Branch, and by rewinding the clock and tracking stellar

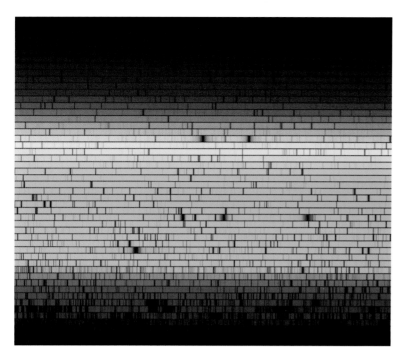

ABOVE The spectrum of the Sun – the dark lines show absorption of light by various elements in the solar atmosphere. *(N. A. Sharp, NOAO/NSO/Kitt Peak FTS/AURA/NSF)*

evolution backwards, they can then figure out how long it has taken these stars to reach these various evolutionary stages, and therefore how old the cluster is.

So, back to the Sun. If the Sun formed inside a molecular cloud with other stars, then it should share several properties with each of those stars. The first, as we have seen, is age. The second is the relative abundance of heavy elements; since the Sun and its siblings came from the same gas cloud, they should all be made of exactly the same ratios of material as that gas cloud. It should be possible to determine this by identifying stars that have a spectrum that is identical, in terms of emission and absorption lines, as the Sun's spectrum. The third property is motion – the Sun's siblings should be orbiting the galaxy in fairly similar orbits. This last property won't be as exact as the first two, since there are things that can perturb the motion of stars, such as the gravity of GMCs, encounters with the denser spiral arms, or perturbations from dwarf galaxies or gas clouds that might collide with the Milky Way (one would imagine that the cloud that impacted the Orion Arm to form Gould's Belt disturbed the trajectories of many stars that got in the way).

However, if astronomers can identify enough of the Sun's birth siblings, it should be possible to track back all their orbits and find a common

point of origin. This would be the Sun's birthplace. These birth siblings could be either in front of the Sun or behind it in their orbit, but so far only one likely candidate has been identified, which is the star HD 162826, currently located 110 light years away in the constellation of Hercules (it's just a bit too faint to be seen with the naked eye, but binoculars and small telescopes will show it at magnitude 6.7). HD 162826 is 15% more massive than the Sun, and shares very similar abundances of heavy elements such as oxygen, carbon, silicon, iron and barium. In a way, identifying the abundance of these elements is like building up a DNA profile, and then seeing which other stars share that same profile.

However, this is just one star, and given its proximity and relative brightness, it was an easy one to identify. As we have seen, though, the initial mass function of molecular clouds trends towards lower mass stars, and these stars are fainter and thus much harder to find. Nevertheless, Simon Portegies Zwart of Leiden University in the Netherlands has estimated that there could be between 10 and 60 of the Sun's siblings within about 300 light years of the Sun, with most of the rest still within about 3,000 light years. Astronomers just have to find them.

Meet the neighbours

The vast majority of the stars around us now may not have been born at the same time and in the same place as the Sun, but for the time being they remain our neighbours. What can the stars – and planets – in the Sun's neighbourhood tell us about the stars in the Milky Way as a whole?

According to the RECONS (Research Consortium On Nearby Stars) project there are 428 stellar objects within ten parsecs – that is, 32.6 light years – of the Sun. Of those objects, four are A-type stars (brighter and hotter than the Sun, with a surface temperature up to 9,700°C/17,492°F); seven are F-type stars (slightly brighter and hotter than the Sun, with a surface temperature up to 7,300°C/ 13,172°F); 19 are G-type stars like our Sun (with a surface temperature up to 5,700°C/ 10,292°F – the Sun's surface temperature is 5,500°C/9,932°F) – 43 are K-type stars (dimmer and cooler than the Sun with a surface temperature up to 5,000°C/9,032°F); and 284 are M-dwarfs (also known as red dwarfs, the dimmest and coolest type of star with a surface temperature no greater than 3,700°C (6,692°F)). Then there are 50 known brown dwarfs, which are failed stars that are not massive enough to generate the temperatures within their cores for hydrogen fusion reactions, and 21 white dwarfs, which are the core remnants of Sun-like stars that expanded into red giants and then puffed off their outer layers as a beautiful but transient planetary nebula. There are none of the hot O- and B-type stars that explode as supernovae, signifying their relative rarity in the Milky Way, while three out of four stars in the closest ten parsecs are red dwarf stars. And the average distance between stars, whatever their type, is about four light years. There's nothing particularly different or unusual about the Sun's neighbourhood; we can expect any given volume ten parsecs wide, chosen at random in the Milky Way, to look pretty similar. The exotic denizens of the galaxy – black holes, giant stars, neutron stars and accreting binaries – are rare indeed.

In the past 25 years, another kind of object has come to the fore: exoplanets. These are worlds that orbit other stars, and statistical analysis of exoplanet discoveries suggests that virtually every star in our galaxy has planets; indeed, they likely have more than one planet if our solar system is anything to go by. Given that there are an estimated 200 billion stars in the Milky Way, there could therefore be trillions of planets, and those are just the ones in orbit around stars – there are potentially hundreds of billions more ejected from their planetary systems via gravitational encounters with their neighbouring planets, and which are now wandering rogue between the stars. Planets form from discs of gas and dust that surround newly born stars, the discs slowly cooling and fragmenting, smaller dust grains coming together to form pebbles, and then increasingly larger bodies until they become big enough to be planets in their own right. For example, we can see such a dusty disc in the process of building planets around the bright star Vega, which is 25 light years distant in the constellation of Lyra (the Harp).

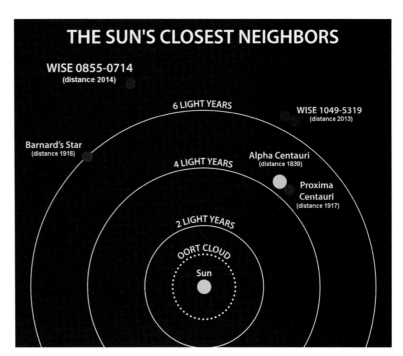

THE SUN'S CLOSEST NEIGHBORS

WISE 0855-0714 (distance 2014)

6 LIGHT YEARS

WISE 1049-5319 (distance 2013)

Barnard's Star (distance 1916)

4 LIGHT YEARS

Alpha Centauri (distance 1839)

Proxima Centauri (distance 1917)

2 LIGHT YEARS

OORT CLOUD

Sun

Over 4,000 exoplanets have been identified, some of which are hundreds or even thousands of light years distant, while others are somewhat closer. Astronomers have discovered exoplanets around the nearby stars Proxima Centauri 4.2 light years away, Barnard's Star 5.9 light years away, and Epsilon Eridani 10.5 light years away, among others. While exoplanets have been discovered ranging in mass from a dozen times the mass of Jupiter to smaller than Mars, the most common type of planet in the Milky Way seems to be what astronomers call 'mini-Neptunes'. In our solar system Neptune is a so-called 'ice giant' that

ABOVE Our Sun's closest stellar neighbours and the year when each of their distances were determined. *(NASA/ Penn State University)*

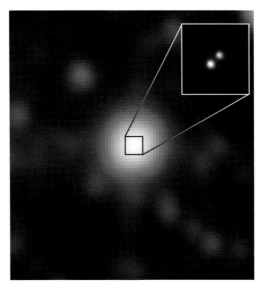

LEFT Brown dwarf binary system Luhman 16 is the closest-known 'failed stars' to Earth, located at a distance of 6.5 light years. *(NASA/JPL/ Gemini Observatory/ AURA/NSF)*

LEFT HD 219134b is a suspected scorching hot rocky planet; one of the closest worlds to our solar system. *(NASA/JPL-Caltech)*

is 17 times more massive and twice as big as Earth, located on the frozen edge of the solar system. In our solar system, Neptune is the fourth largest and most massive planet, ranking behind Jupiter, Saturn and Uranus. However, after Neptune there is a large gap until you get to Earth.

The mini-Neptunes help fill this gap, weighing in with masses between that of Earth

LEFT The six-Earth-mass alien world Gliese 667 Cb orbits its nearby low-mass host star at a distance of 1/20th of the Earth–Sun distance. *(ESO/L. Calçada)*

BELOW LEFT An artist's impression of the sunset on Gliese 667 Cc – the planet orbits in the habitable zone of its star Gliese C in the Gliese 667 system and is one of the closest exoplanets to Earth at almost 24 light years. *(ESO/L. Calçada)*

BELOW Proxima Centauri is our closest stellar neighbour and is part of a triple star system. Its two companions are Alpha Centauri A and B. *(ESA/Hubble & NASA)*

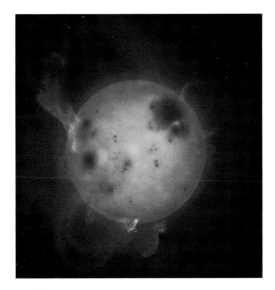

and Neptune, and diameters somewhere in between too. Most crucially, however, is the fact that they are mostly quite close to their stars, where temperatures are far hotter than the minus 214°C (minus 353.2°F) of frozen giant Neptune. In our solar system, Neptune is rich in volatiles (substances with low boiling points) such as water and carbon dioxide, but in the frigid temperatures these become ice crystals. Closer to their stars, however, the mini-Neptunes melt these ice crystals and then sublimate them, such that their atmospheres are rich in water vapour, possibly with an ocean below the clouds.

As of the time of writing, no truly Earth-

ABOVE LEFT Proxima Centauri is a flare star – it is prone to dramatic and random changes in brightness. *(University of Warwick/Mark Garlick)*

ABOVE An artist's impression of the surface of planet Proxima b orbiting the red dwarf star Proxima Centauri. *(ESO/M. Kornmesser)*

like planet has yet been discovered. To be Earth-like, an exoplanet would need to have a nitrogen-oxygen atmosphere, to be at just the right distance from its star to be at the appropriate temperature for liquid water to exist on its surface, and to carry the potential to possess a habitable environment. So far, Earth remains unique in our Milky Way galaxy.

LEFT Epsilon Eridani b has yet to be confirmed, but if it exists it will be ten light years away in the constellation of Eridanus (the River). *(NASA, ESA, G. Bacon)*

Chapter Five

Unlocking the Milky Way's secrets

For most of human history, the night sky, and the Milky Way that streams through it, were a mystery. Ancient cultures invented stories; myths to make sense of the stars with. To the Mayans, it was the 'World Tree' that provided a path from the underworld to heaven. To the ancient Chinese, the Milky Way was the Silver River, a line drawn in the sky by the Queen of Heaven to separate the fairy Zhi Nu (represented by the bright star Vega) from her lover, the cowherd Niu Lang (the star Altair), who had eloped.

OPPOSITE The Gaia spacecraft is designed to create a three-dimensional map of the Milky Way. It was launched in 2013. *(ESA–D. Ducros, 2013)*

Of course, it is the Ancient Greeks whose myths dominate the night sky from a Western perspective. As always, the Greek myths were not shy on graphic details. The modern-day name, Milky Way, is a literal translation from the Latin Via Lactea. The story goes that the hero Hercules was born to an immortal father – Zeus – and a mortal mother, Alcmene. Zeus had plans for Hercules, however, and to give him the super-strength that Hercules is today famed for, his father surreptitiously had him feed at the breast of the Goddess Hera while Hera slept. One night Hera awoke to find the baby Hercules at her chest, and tore him away, splashing milk across the sky.

Not all the Ancient Greeks chose the mythological explanation. Aristotle thought that the Milky Way was a product of some 'fiery exhalation' in Earth's atmosphere originating from a great number of stars. Other pre-Renaissance philosophers shared a similar view, but it was only following the invention of the telescope that we came to gradually understand what the Milky Way really is.

As you'll remember from Chapter one, Galileo was the first to conclusively show that the Milky Way is made from stars too faint to resolve individually with the naked eye. In 1785 William Herschel tried mapping the Milky Way to figure out its true shape. Then in the 20th century Edwin Hubble discovered that there were other galaxies beyond our Milky Way, and that our Milky Way galaxy quite probably looked like the other galaxies. Ever since, our knowledge of the Milky Way has grown cumulatively, as we have also moved beyond visible light to observe at other wavelengths, and the key to this increasing knowledge has been the increasing capabilities of our telescopic technology. So let's look at some of the telescopic marvels that are helping to teach us more about our own galaxy.

The Hubble Space Telescope's view

Let's begin with perhaps the most famous telescope of all time – the Hubble Space Telescope. Earth's atmosphere is somewhat debilitating; turbulence caused by atmospheric motions and thermal currents affects what astronomers refer to as 'seeing' – it's the twinkling of the stars in the sky that muddies the view. One way to obtain a clearer view is to get above the atmosphere by placing a telescope in space.

The 13m 42ft 8in) long Hubble Space Telescope launched on board the Space Shuttle *Discovery* in April 1990. Named after Edwin Hubble, who did so much to transform our understanding of the Milky Way and the universe, the space telescope was originally beset by problems, with budget overruns bringing its cost at launch to $4.7 billion (£3.8 billion). Then, once in orbit, 569km (353 miles) above our heads, disaster struck. The first images from Hubble were blurry. A thorough investigation promptly discovered the cause: the telescope's 2.4m (7ft 10in) diameter mirror had been perfectly polished to the wrong specification. The difference between the shape of the mirror and what it should have been was tiny, just 2.2 microns (millionths of a metre), but this was enough to result in the telescope not being able to focus properly. The optical term for this malady is spherical aberration, but in everyday terms, the Hubble Space Telescope was effectively short-sighted.

So, what does a short-sighted person do? They wear glasses, and indeed this was the solution to Hubble's blurry eyesight. During a daring space mission in December 1993 involving five spacewalks, astronauts on board the Space Shuttle *Endeavour* installed a device, called COSTAR, the Corrective Optics Space Telescope Axial Replacement, which was placed in the optical train of the telescope. It contained two mirrors, one of which is ground to cancel out the spherical aberration of the primary mirror – in effect, giving Hubble glasses. The plan worked a treat, with the new images taken through COSTAR having the pinpoint sharpness we've come to expect from Hubble.

In later years, COSTAR became redundant. Hubble has space for five scientific instruments and one, the High Speed Photometer, had to be removed to make way for COSTAR. However, the mission to fix Hubble's vision was only the first of the space telescope's servicing missions – in 1997, 1999, 2002 and 2009 astronauts returned to Hubble, grasping it in their space shuttle's robotic arm and pulling it into the shuttle's cargo bay, with doors open to space, to replace or fix instruments that needed upgrading or that had broken down. These new instruments were built to account for the spherical aberration themselves, so COSTAR was removed and its space made available to another scientific instrument.

With the retirement of the space shuttle in 2011, no more servicing missions are possible. Hubble has been left with a functioning quartet of the Wide Field Camera 3 (WFC3), the Space Telescope Imaging Spectrograph (STIS), the Cosmic Origins Spectrograph (COS) and the Advanced Camera for Surveys (ACS) with which to observe the universe. It also still carries the Near-Infrared Camera and Multi-Object Spectrometer (NICMOS), but this is currently deactivated since the WFC3 can make pretty much the same observations.

Hubble does a lot of unheralded work – things like spectroscopic observations and deep surveys that are vital for astronomical research, but aren't about the whizz-bang colourful images that the space telescope is most famous for. Hubble really has brought to us parts of the Milky Way galaxy in wonderful high-definition. In 1995 it took possibly the most famous astronomical image ever, of the Pillars of Creation. These are three towering columns of gas, light years long, that are forming stars within the Eagle Nebula (see Chapter 6 for more details). It was the perfect example of how Hubble can reveal to us the intricate

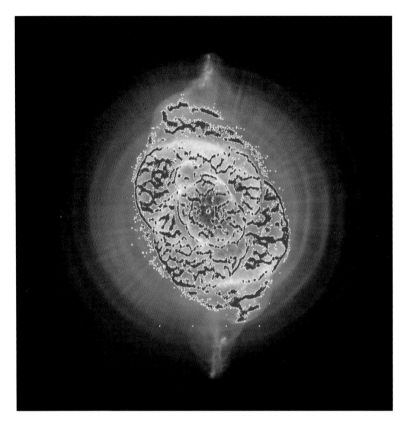

workings of star formation within the Milky Way, showing details never seen before, thanks to its angular resolution of 0.1 arcseconds (0.0000278 degrees). It is also teaching us new things about star-death.

Although Hubble is primarily an optical telescope, it has the ability to observe in near-

ABOVE The Cat's Eye Nebula (NGC 6543). The Hubble Space Telescope shot the background image, while Gaia's blue smudges point to over 84,000 detections of gaseous filaments. *(NASA/ESA/HEIC/ The Hubble Heritage Team/STScI/AURA (background image); ESA/Gaia/DPAC/UB/ IEEC (blue points))*

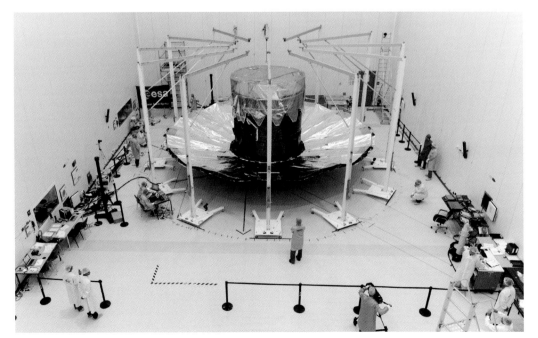

LEFT Gaia during testing in Kourou, French Guiana. *(ESA–M. Pedoussaut)*

infrared and ultraviolet wavelengths. Infrared, in particular, is helpful for seeing through the dust around the galactic centre. For example, Hubble's NICMOS instrument, combined with longer-wavelength infrared data from NASA's Spitzer Space Telescope, was able to produce an unrivalled vista of the galactic centre. It showed the giant star clusters that exist there, known as the Arches, the Quintuplet and Central clusters. It had been thought that these clusters, born from recent bursts of star formation, contained all the young, massive, luminous stars in the galactic centre, but NICMOS was able to identify massive stars that had gone rogue. Perhaps they formed in another cluster that was pulled apart by immense gravitational forces in the area.

Hubble has also studied less massive stars just outside the galactic centre, in the Milky Way's pseudobulge. Many of Hubble's discoveries don't come from targeted observations, but from data plucked from its large surveys across swathes of the sky, and studies of the pseudobulge can take advantage of this large amount of data on thousands of stars. In 2018 a team led by Will Clarkson of the University of Michigan-Dearborn mined data from two of Hubble's surveys – the WFC3 Galactic Bulge Treasury Program and the Sagittarius Window Eclipsing Extrasolar Planet Search, performed by the WFC3 and ACS – to find that the pseudobulge contains many different populations of stars. They found that stars with higher metallicity (and which are therefore younger) have less chaotic and faster orbits, whereas stars with more primitive chemistry (and which are

therefore older) have more disordered orbits and are travelling more slowly. This strengthens the argument that the pseudobulge formed from more gradual secular processes, rather than forming all at once 12 to 13 billion years ago, as had originally been thought.

On a broader scale, Hubble, in conjunction with the Gaia satellite, has also been able to measure the mass of the Milky Way. The pair of orbiting observatories accurately measured the motions of 46 globular clusters. The more massive the Milky Way is, the faster they should orbit and, based on their velocities, Hubble and Gaia were able to pin down the mass of the Milky Way as 1.5 trillion solar masses, narrowing it down from previous estimates that ranged between 500 billion and three trillion solar masses.

Mapping the Milky Way

One of the big obstacles that lies between us and understanding the Milky Way better is our lack of ability to map it accurately. Measuring distances in space is very difficult without some kind of distance marker, such as a standard candle like a Cepheid variable. This is where the European Space Agency's Gaia spacecraft – and before that its predecessor, Hipparcos – comes in.

Gaia is an astrometric mission, meaning that it is designed to measure the positions and motions of over a billion stars with an incredibly high accuracy of 24 microarcseconds – making it 200 times more precise than Hipparcos was able to achieve in 1989. In 2018, Gaia's mission scientists presented their second data release that acts as a three-dimensional map of all those stars – the first map of its kind. By knowing the motions and positions of those stars, it is possible to picture where they will be in the future, or rewind the clock and calculate where the stars were in the past.

Scientists call it galactic archaeology. An initial release of data from Gaia in 2016 gave a taster of what was to come in the full-blown second data release in April 2018, which was based on 22 months of observations and which featured information on the positions and brightnesses of 1,692,919,135 (1.69 billion) stars, parallax distance measurements and proper motions (i.e. the star's movement

BELOW A strange collection of stars uncovered in this image from Gaia. With a different chemical composition, they could have come from another galaxy that crashed into our young Milky Way – astronomers have dubbed the intruder Gaia-Enceladus. *(ESA/Gaia/DPAC; A. Helmi et al)*

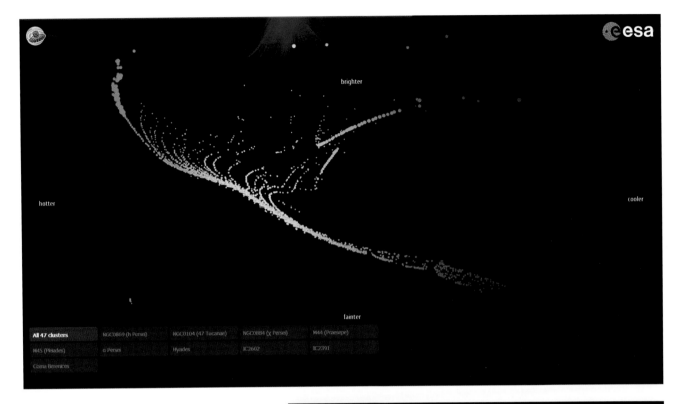

ABOVE Gaia's Hertzsprung–Russell diagram of the stars in our Milky Way. *(ESA/Gaia/DPAC)*

RIGHT The Gaia spacecraft produces the richest star catalogue to date – the map features measurements of almost 1.7 billion stars to the highest precision to date. *(ESA/Gaia)*

against the fixed frame of the background sky) for 1,331,909,727 (1.33 billion) stars, the surface temperatures, and hence colours, of 161,497,595 (161 million) stars, and the amount of interstellar dust along the line of sight of 87,733,672 (87.7 million) stars, which is detected by the amount by which it absorbs and reddens starlight and can therefore inform scientists about the distribution of interstellar

RIGHT With the oval representing the celestial sphere, this is how the Gaia spacecraft scans the sky. The colours reveal how often the portions of the sky were observed, with blue depicting the most scanned regions and the lighter colours being the least studied. *(ESA/Gaia/DPAC; acknowledgement: B. Holl (University of Geneva, Switzerland) on behalf of DPAC)*

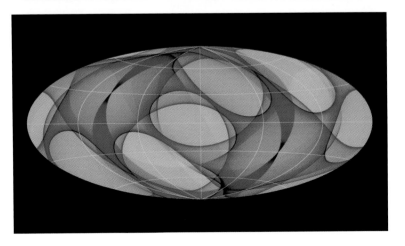

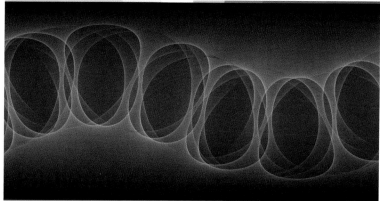

ABOVE Gaia's map of the stellar density of our galaxy. The Large Magellanic Cloud (LMC), Small Magellanic Cloud (SMC), NGC 104 (left of the SMC), NGC 6205 (left of the galactic core) and NGC 7078 (below galactic core) are also visible. *(ESA/Gaia/Edmund Serpell)*

LEFT Real data that shows the viewing orientation of Gaia – the spacecraft rotates slowly, sweeping its pair of telescopes to create four complete rotations per day. The telescope's spin axis also changes position around the Sun on a 63-day period. *(ESA/Gaia)*

dust in the Milky Way. With all this data, Gaia scientists have also created the most intricate Hertzsprung–Russell diagram of the stars in the Milky Way that has ever been assembled. In all, Gaia has provided detailed data on almost 1% of all the stars in our galaxy.

Gaia is also making new discoveries. Scrutinising the spacecraft's data, astronomer

LEFT Inside the control room at the European Space Agency in Darmstadt, Germany. Here, the operations teams are led by Spacecraft Operations Manager David Milligan (pictured). *(ESA)*

Sergey Koposov of Carnegie Mellon University, Pennsylvania, spotted a concentration of stars very close in the sky to the brightest star in the sky, Sirius. Hidden in the star's glare, the cluster – subsequently known as Gaia 1 – is remarkable. It may be close in the sky to Sirius, which is 8.6 light years from us, but in reality Gaia 1 is 15,000 light years further away. It's pretty big for an open star cluster, containing a few thousand solar masses spread across a region 30 light years in size. This might explain why, at an age of three billion years, Gaia 1 is

still holding together – normally open clusters drift apart after a few hundred million years, but in Gaia 1 the force of gravity may be keeping the stars together. It's also on a peculiar orbit, climbing up to 3,000 light years above the plane of the Milky Way's disc before diving back in. The mystery is that these trips in and out of the galactic disc should see Gaia 1's stars whittled down and stripped away by gravitational tidal forces from the disc, but after three billion years of this, it does not seem to have happened. Furthermore, no other open star cluster has

ABOVE The Spitzer Space Telescope has been observing the galaxy since its launch in 2003. The mission will come to an end in 2020. *(NASA/JPL-Caltech)*

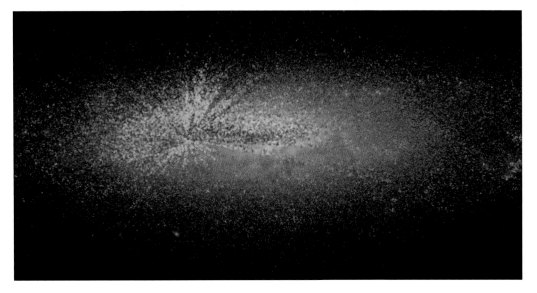

LEFT Coloured dots superimposed on a concept of the Milky Way reveal stars that were made when our galaxy was young (red) and those that formed when it matured (blue). The data is from the Sloan Digital Survey. *(G. Stinson (MPIA))*

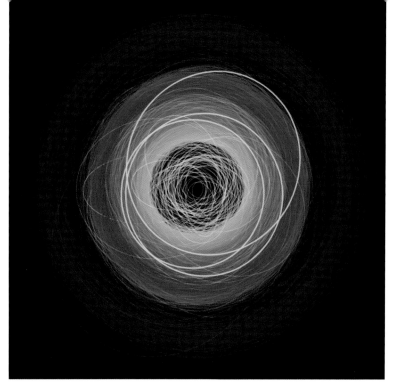

ABOVE As well as scanning the sky to chart the stars of the Milky Way, the Gaia spacecraft regularly observes asteroids in our solar system. The orbits of more than 14,000 known asteroids are shown in this image, with the Sun at the centre. *(ESA/Gaia/DPAC)*

ABOVE RIGHT This ground radio telescope regularly communicates with the Gaia spacecraft. *(ESA/S. Halté)*

RIGHT A three-dimensional map of the hottest, brightest and most massive stars in our galaxy's disc. The data from Gaia is based on 400,000 stars within less than 10,000 light years from the Sun. *(Galaxy Map/K. Jardine)*

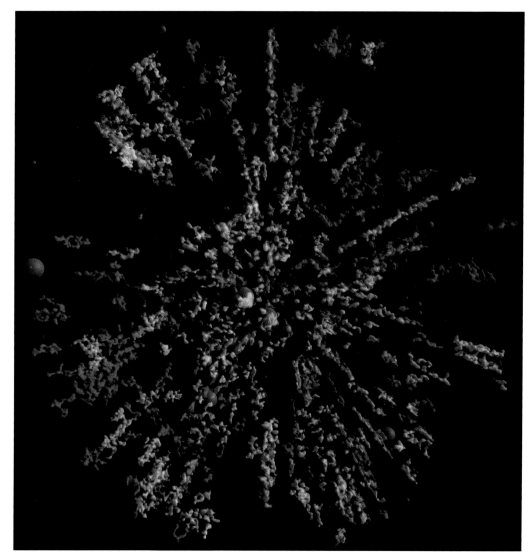

been found on such a steep orbit around the galaxy, leading astronomers to wonder about Gaia 1's origins.

Gaia has also found evidence of other galaxies interacting with the Milky Way. Most recent (between 200 million and 1 billion years ago) is a close encounter with what was probably the Sagittarius dwarf galaxy, the gravitational tides which produced ripples in the Milky Way's disc, reflected in the motions of stars that Gaia was able to detect and measure. Much more ancient is the evidence for the young Milky Way colliding and merging with another galaxy that was about the size of the Magellanic Clouds. This merger took place more than 10 billion years ago, before the formation of the Milky Way's thick disc – in fact, the collision has spurred on the formation of the disc. The galaxy that merged with the Milky Way has been posthumously named Gaia-Enceladus, for the giant of Ancient Greek mythology that was the child of Gaia, who was the mother goddess of Earth, and Uranus, who was the god of the sky. Gaia's evidence for this merger comes in the form of a group of 30,000 stars in the Milky Way's halo that have anomalous motions and chemical compositions that imply that they're not originally from around these parts, but were integrated into the Milky Way when their parent galaxy was subsumed by our galaxy. In addition, 13 globular clusters also have motions that suggest they originated in Gaia-Enceladus.

ABOVE Gaia uncovered a lull in the speeds of stars in the Milky Way, suggesting that the cannibalisation of the Sagittarius dwarf galaxy could be to blame. *(ESA)*

BELOW A so-called Gaia-Enceladus galaxy supposedly merged with our galaxy during its early formation some 10 billion years ago. Its debris can be found throughout the Milky Way. *(ESA/Koppelman, Villalobos and Helmi)*

VISTA: Surveying the galaxy from Earth

Located at the European Southern Observatory on Cerro Paranal in Chile's Atacama Desert, the UK-built Visible and Infrared Survey Telescope for Astronomy, or VISTA, is currently the world's largest survey telescope with a primary mirror 4.1m (13ft 6in) across, although this will soon be supplanted in the early 2020s by the 8.4m (27ft 7in) Large Synoptic Survey Telescope (LSST) on the neighbouring peak of Cerro Pachón.

VISTA's principal work is spent conducting six major surveys of the night sky in the Southern Hemisphere. From its perch atop Cerro Paranal, which is 2,635m (8,645ft), at the foot of the Andes mountains, VISTA is above much of the turbulent atmosphere and water vapour, giving it a great view of the centre of the Milky Way in Sagittarius.

Several of the surveys relate to extragalactic activities. For example, the VISTA Kilo-Degree Infrared Galaxy Survey (VIKING) is scouring

ABOVE The Visible and Infrared Survey Telescope for Astronomy (VISTA). *(A. Tyndall/ESO)*

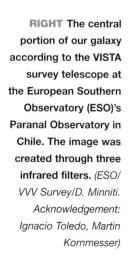

RIGHT The central portion of our galaxy according to the VISTA survey telescope at the European Southern Observatory (ESO)'s Paranal Observatory in Chile. The image was created through three infrared filters. *(ESO/VVV Survey/D. Minniti. Acknowledgement: Ignacio Toledo, Martin Kornmesser)*

1,500 square degrees of the sky to observe the colours of galaxies as a way of measuring their distance via the technique of photometric red shift – the redder a galaxy is overall, the more likely that this is an effect of red shifting – the stretching of a galaxy's light with the expansion of the universe – and hence the easier it is to estimate its distance. Meanwhile, the UltraVISTA survey is mining a smaller region of sky, with an area of just 0.73 square degrees, to make deep images capable of detecting some of the first galaxies to form after the Big Bang, when the universe was just a few billion years old. Then there's VIDEO, the VISTA Deep Extragalactic Observations Survey, which is focused on studying active galaxies, the formation and development of galaxy clusters, and some of the massive elliptical galaxies that reside within them.

A little closer to home, the VISTA Magellanic Survey (VMC) is observing the Large and Small Magellanic Clouds, but the two that are most relevant to our Milky Way studies are the VISTA Hemisphere Survey (VHA) and the VISTA Variables in the Via Lactea (VVV) survey. The former is imaging the entire southern half of the celestial sphere to study low-mass stars in the solar neighbourhood and to better understand the galaxy's merger history that has seen it incorporate stars from other galaxies that the Milky Way has cannibalised, while the latter is studying 520 square degrees of sky centred on the galactic centre, searching for variable stars. As we have seen, variable stars such as Cepheid variables are incredibly important to astronomers, because their period-luminosity relation (as a brief reminder, the more slowly they pulsate, the more luminous they are at peak brightness) can be used to measure their distance. In this way, the VVV survey is able to build a three-dimensional map of the stars in the pseudobulge at the core of the Milky Way.

The VVV survey has made some surprising discoveries. In 2015 astronomers led by Istvan Dékány of the Pontificia Universidad Católica de Chile analysed 655 Cepheid variable candidates detected by VVV. Cepheid variables tend to come in two groups – there are those that are very old, smaller stars, and those that are younger, hotter and brighter. We tend to find the former in the bulge and the latter in

THE POWER OF VIRCAM

VISTA is a powerful astronomical tool, as evidenced by its impressive infrared camera, called the VISTA InfraRed CAMera (VIRCAM). This three-tonne chunk of equipment is provided a 1.65-degree field of view by VISTA's large mirror, and contains 16 infrared sensors totalling 67 megapixels. It operates in near-infrared wavelengths (0.7–5 microns) – those wavelengths closest to the visible red part of the spectrum, and which are not badly attenuated by water vapour in Earth's atmosphere, unlike mid- and far-infrared wavelengths (5–40 microns and 40–350 microns respectively). In particular, VIRCAM operates with several broadband filters (meaning they observe across a wide range of wavelengths) and three narrowband filters that detect light at the specific wavelengths of 0.98, 0.99 and 1.18 microns respectively.

BELOW Observations of the Large Magellanic Cloud (LMC), one of the Milky Way's satellites from ESA's Gaia satellite – it depicts the total amount of radiation and the rotating of several million stars. *(ESA/Gaia/DPAC)*

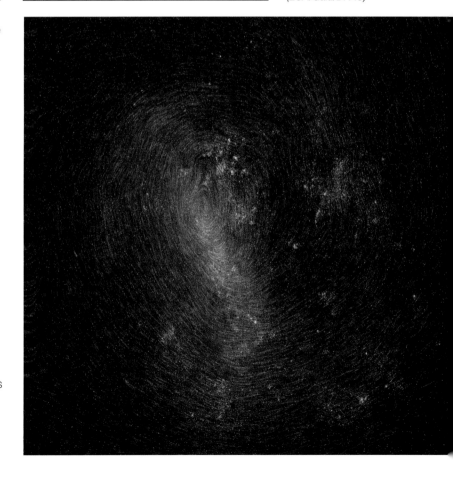

the stellar thin disc. However, Dékány's group were able to identify 35 of the Cepheids as belonging to the younger group; in fact, all of them are younger than 100 million years, with the youngest so far discovered being just 25 million years old.

If the Milky Way's bulge was a typical bulge, we would expect its stars to be older (with the exception of in the immediate vicinity of the supermassive black hole Sagittarius A*, where there is lots of gas and frenzied star formation). The presence of these young Cepheids implies that the pseudobulge is receiving a fresh supply of young stars, likely transported there by the galaxy's central bar.

However, the discovery of the Cepheids per se was not the main discovery that Dékány's group made. Instead, it was the discovery that these young Cepheid stars, alongside other young stars that are not Cepheids, inhabit a slim disc-like structure at the heart of the galaxy – a new component of the Milky Way, never before seen because it is hidden behind dense clouds of interstellar dust that congregate in the centre of the galaxy and through which only longer wavelengths of light – infrared, radio – can pass.

VVV hasn't only been finding young Cepheids in the galactic centre. As well as the older Cepheids, VVV has also identified a dozen or so RR Lyrae variables. These stars have period-luminosity relationships just like Cepheids do, except they tend to have shorter periods and their stars are of different spectral classes and compositions. More crucially, until now they had been found exclusively in the Milky Way's halo, specifically within globular clusters. So what are they doing in the Milky Way's bulge?

Their presence could be telling us something about how the inner 400 light years of the galaxy formed. Perceived wisdom has been that the central stellar bulge, which was the first part of the galaxy to form 12–13 billion years ago, had quickly assembled from the rapid accretion of primordial gas clouds at the core of the dark matter halo in the cosmic web. However, the presence of the RR Lyrae stars suggests something different – that the central nuclear bulge was formed by the merger of many ancient globular clusters that contained these RR Lyrae stars. It remains to be seen if this is the explanation, and whether the gas accretion model can be completely ruled out, or whether

BELOW A bird's-eye view of the Very Large Telescope (VLT). *(ESO/G.Hüdepohl)*

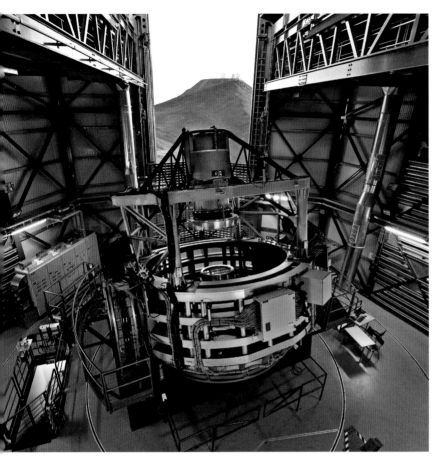

the gas accretion and globular cluster mergers acted in unison to form the hub of the Milky Way all those years ago.

The mountains that rise out of Chile's Atacama Desert are a hotspot for astronomical observatories. VISTA is only 1.5km (4,920ft) from the European Southern Observatory's Very Large Telescope (VLT), which consists of four distinct 8.2m (27ft) telescopes that can operate individually or together as the VLT Interferometer. It too has been used to probe the galactic centre. For example, it has the resolution required to have allowed astronomers led by Reinhard Genzel of the Max Planck Institute for Extraterrestrial Physics in Germany to have tracked the motions of stars as they get as close as 17.9 billion km (11.1 billion miles) to the supermassive black hole. Although the Hubble Space Telescope has been placed into orbit to achieve the crystal clear views required for such accurate observations, the ground-based VLT is able to achieve such views through the use of adaptive optics that are able to counteract atmospheric seeing. An artificial guide star is created in the sky by a laser beam shining up

ABOVE With the Very Large Telescope visible on a nearby mountain, here stands the VISTA telescope. It is the biggest survey telescope on the globe with a primary mirror size of 4.1m (13ft 6in). It views the universe through near-infrared wavelengths.
(G. Hüdepohl/ESO)

RIGHT Three of the Very Large Telescope (VLT)'s four Unit Telescopes get ready for a night of observing the Milky Way from their location on top of Cerro Paranal, Chile.
(ESO/G.Hüdepohl)

from the observatory. Computers monitor how the appearance of this guide star fluctuates in the atmospheric turbulence, and then control actuators that minutely adjust the shape of the mirror in real time to correct for the turbulence in the atmosphere. Such technology is commonplace now on all large professional observatories.

ALMA: Making the invisible visible

One breed of telescope that doesn't have to worry about the seeing conditions is those that operate at very long wavelengths, in the submillimetre, microwave and radio regimes.

The Atacama Large Millimeter/submillimeter Array, better known as ALMA, is the most expensive ground-based telescope built up until the year 2019, costing $1.4 billion, although at least one (the Square Kilometre Array) and possibly several new telescopes (the Extremely Large Telescope and the Thirty Meter Telescope), set to come into operation in the 2020s, will cost more.

What all this money buys is a lot of scientific discoveries. Consisting of 66 moveable radio dishes – 50 of which are 12m (39ft 4in) in diameter, with the remaining dishes being 7m (22ft 11in) – on the Chajnantor Plateau in Chile's Atacama Desert, not too far from VISTA and the VLT, ALMA specialises in astrochemistry. Millimetre wave light is commonly emitted by molecules in space, and ALMA is discovering all manner of molecules and compounds. It has spotted methyl isocyanate, which is a complex organic (i.e. carbon containing) molecule, in the multiple star-forming IRAS 16293-2422 system, 400 light years away in the Rho Ophiuchi star-forming region.

It was the first time that such a complex organic had been found around a young Sun-like star. Methyl isocyanate (chemical formula MIC) is a toxic substance, but it belongs to a class of organics that are complex enough to bond to form amino acids, which in turn can form peptides and proteins that are vital for the mechanics of life as we know it. In other words, astro-chemists are now starting to find the first step on the way to the formation of the materials for life around other stars, which is

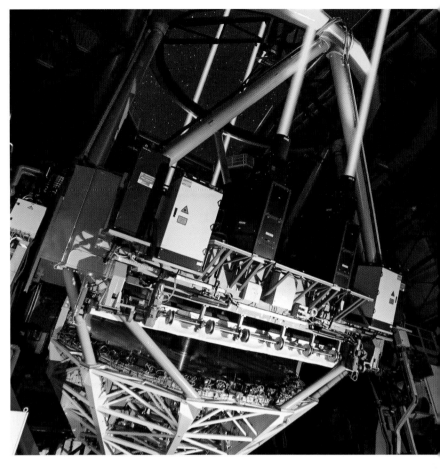

great news for astrobiologists. Other complex organic molecules that ALMA has found around young stars include methyl alcohol (a.k.a. methanol, CH_3OH) in the protoplanetary disc surrounding the star TW Hydrae 170 light years away in the constellation of Hydra, and methyl cyanide (CH_3CN) and hydrogen cyanide (HCN) in the outermost cold reaches of the protoplanetary disc around the star MWC 480, which is 450 light years away in the constellation of Taurus. One concludes from ALMA's findings that the Milky Way must be absolutely teeming with the basic ingredients for life, although whether it has formed life anywhere else in the universe bar Earth is a mystery that we still don't know the answer to!

ALMA has also detected where planets are forming on which such hypothetical life could exist. It has been able to clearly observe the dust discs around young stars, and found patterns in them – gaps, rings and ripples, which are understood to be caused by planets in the process of forming, sweeping up the dust to clear out gaps while their gravity

ABOVE The Unit Telescope 4 (UT4) of the Very Large Telescope (VLT) looks up to the stars of our galaxy. *(ESO/F. Kamphues)*

ABOVE At 5,000m (16,400ft) above sea level on the Chajnantor Plateau, here stands the ALMA antennas – there are 19 radio dishes in total. *(ALMA (ESO/NAOJ/ NRAO)/W. Garnier (ALMA))*

perturbs dust elsewhere in the disc. ALMA has even been able to probe even earlier in a young star's life, before the protoplanetary disc and even the star itself has formed. It's witnessed dark cores of dense gas and dust called Infrared Dark Clouds, which are currently devoid of stars but are on the brink of forming them through gravitational collapse. It's a precious look into the very first stage of star formation in the galaxy.

Another mysterious aspect of star formation in the galaxy is what ALMA has discovered at the centre of the Milky Way. We've already established in this book that the supermassive black hole, Sagittarius A*, has immense gravitational fields that was tides over anything nearby, pulling objects such as planets, stars and gas clouds apart if they get too close. We also know that the centre of the galaxy is a haven for star formation, but these are in dense, gravitationally bound clusters that can resist the black hole's overtures.

In the past, radio observations have found proplyds in the galactic centre, which are very young stars cocooned inside their own womb of gas and dust that have also been found aplenty elsewhere in the Orion Nebula (see Chapter six). In the galactic centre they

are thought to be dense enough and massive enough to resist the black hole's gravity at their distance. However, ALMA has discovered 11 young, low-mass stars in the process of forming within just three light years of Sagittarius A*. This shouldn't be possible, because the mass of the young stars and the clouds of gas that are forming them should not be great enough to resist the gravitational tides at that distance. It's possible that the black hole itself is aiding their formation by causing the formation of jets or radiation pressure from material accreting onto the black hole that then usher gas together to form the stars. These young stars are estimated to be just 6,000 years old, and ALMA was able to identify them by detecting molecular gas flowing onto them and building up their mass.

However, neither ALMA nor radio telescopes can detect molecular hydrogen, because it is too cold to emit light. It's really only in active star-forming regions illuminated by hot young stars that we can see the molecular hydrogen. However, for every 10,000 molecules of hydrogen in the Milky Way, it is thought that there is one molecule of carbon monoxide (CO), and CO is detectable by ALMA, so it uses it as a proxy to trace the location and motion of molecular hydrogen in the galaxy.

Radio observations of our galaxy

Although molecular hydrogen, made of several hydrogen atoms, can easily escape detection, the same cannot be said for atomic, neutral hydrogen gas, i.e. gas made purely from individual atoms of hydrogen that are not bonded together into molecules. Atomic hydrogen radiates strongly at a wavelength of 21cm (which equates to a frequency of 1,420 Megahertz (MHz)), smack bang in the radio part of the electromagnetic spectrum.

It was realised in the 1930s and 1940s that atomic hydrogen would have a strong signature at radio wavelengths, and 21cm radio waves from interstellar hydrogen was detected for the first time in 1951, by Harold Ewen and Edward Purcell at Harvard University using their purpose-built 'horn antenna' – a primitive radio telescope shaped a bit like a funnel and which cost just $500 (£400).

Realising that an entirely new frontier of radio astronomy was now available to them, astronomers began to map the atomic hydrogen gas in the Milky Way, and soon found that it traces out a spiral pattern. Although it had been predicted that the Milky Way had spiral arms like other spiral galaxies, this was the very first time that they had been detected!

Radio telescopes are the largest telescopes on the planet. They have to be because of the long wavelengths that radio has. The public most commonly think of radio telescopes as giant dishes staring up at the sky, such as the 76m (~256ft) Lovell Telescope at Jodrell Bank, and the 100m (328ft) dishes of the Robert C. Byrd Green Bank Telescope in the USA and the Effelsberg Telescope in Germany. The largest radio telescopes are the famous 305m (1000ft) Arecibo radio telescope in Puerto Rico, and the new Five-hundred metre Aperture Spherical Telescope, or FAST, in China.

However, radio telescopes can also be networked together to create a giant telescope made from many smaller telescopes, through a process called interferometry. Just like ALMA, there are observatories with forests of telescopes, such as the Very Large Array (VLA) of 27 radio dishes and the Allen Telescope Array of 42 dishes, both in the USA. The arrays have the advantage of collectively having a large collecting area without the expense or technical challenge of having to build a giant dish. These arrays can also be linked up with other radio telescopes, not just in the same country, but around the world. The Very Long Baseline Array (VLBA), for instance, which spans the United States from Hawaii to New Hampshire, consists of ten 25m (82ft) dishes. The VLBA can also be joined up with the 305m (1000ft) Arecibo radio telescope in Puerto Rico, as well as Green Bank, the VLA and Effelsberg, to for the High Sensitivity Array, with a maximum baseline (i.e. the maximum distance between two telescopes) of 8,611km (5,351 miles). The greater the baseline, the greater the angular resolution, and the more telescopes that you have in the network, the greater the overall collecting area and the higher the sensitivity, i.e. the ability to detect fainter radio emissions.

Over the years, the mapping of our Milky Way with radio telescopes has evolved, and astronomers don't just rely on hydrogen 21cm emission anymore. For example, astronomers can now trace water and methanol molecules in the galaxy thanks to the Masers they create. Masers – that is, Microwave Amplification by Stimulated Emission of Radiation – are basically the natural microwave equivalent of lasers. Clouds containing concentrations of these molecules absorb photons of light. One molecule will then re-emit the photon as a microwave, prompting its neighbouring molecules to do the same, and all of a sudden there's a cascade of photons being absorbed and re-emitted, amplifying them until they are powerful enough to be detected. So modern radio astronomers use these masers to better trace structure in the Milky Way's spiral arms. The motion of the hydrogen gas and masers can therefore also tell us about the rotation of the Milky Way, by the way the radio emission is red or blue shifted as it moves. And since radio waves can pass through gas and dust, they give us much deeper views into the Milky Way than optical light, or even some infrared wavelengths, can.

Into the galaxy's heart

One particular astronomy experiment that is making history is the Event Horizon Telescope (EHT). Led by Shep Doeleman of Harvard University, the project is to directly image the event horizon around the supermassive black hole, Sagittarius A*, at the centre of the Milky Way. The event horizon is the radius from the black hole inside which not even light can escape the great gravitational grasp of the black hole. In order to image the event horizon, Doeleman's team of 200 scientists from over 60 institutions around the world have assembled the largest telescope on Earth. It consists of eight submillimetre observatories dotted around the world: ALMA

and the Atacama Pathfinder Experiment in Chile, the James Clerk Maxwell Telescope and the Submillimeter Array in Hawaii, the Submillimeter Telescope in Arizona, the Large Millimeter Telescope in Mexico, the IRAM 30 metre telescope in Spain and the South Pole Telescope in Antarctica.

ALMA and the South Pole Telescope were the two most vital pieces of the jigsaw. ALMA is the most powerful member of the Event Horizon Telescope, while the South Pole Telescope was vital for increasing the baseline to achieve the necessary angular resolution, which is 20 microarcseconds, the equivalent of reading a newspaper in New York through your telescope in Paris. During simultaneous observing campaigns in April 2017 (each telescope also has regular astronomy work to do, so it can only devote itself to the EHT for a short time each year), the EHT attempted to image both Sagittarius A*, and the supermassive black hole in the elliptical galaxy M87, in the Virgo Cluster. Although M87 is 54 million light years away, compared to the 26,000 light years between Earth and Sagittarius A*, the two black holes appear a similar size on the sky, because M87's

BELOW The locations of some of the telescopes that form the Event Horizon Telescope (EHT). An array of 11 observatories all over the world comprise the virtual instrument: ALMA, APEX, the Greenland Telescope, the IRAM 30-meter Telescope, the IRAM NOEMA Observatory, the Kitt Peak Telescope, the James Clerk Maxwell Telescope, the Large Millimeter Telescope Alfonso Serrano, the Submillimeter Array, the Submillimeter Telescope and the South Pole Telescope. *(ESO/ L. Calçada)*

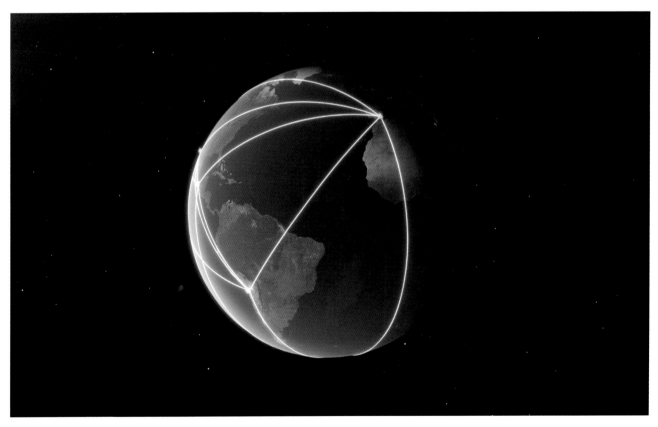

black hole is far more massive than Sagittarius A*, and its event horizon is far wider.

In April 2019, the EHT team revealed their first image of the black hole inside M87. This historic image showed a ring of emission encircling the dark of the black hole itself. The ring is called the photon ring, and is where accreting material is piling up around the black hole and growing hot. Because it is so close to the black hole, the photons are actually travelling all around the black hole, following how its gravity is curving space-time, before leaving and heading towards us. Meanwhile, the dark area inside the bright ring is called the 'shadow' of the black hole, and is the gravitationally lensed image (i.e. the light is bent and magnified by the extreme curvature of space, like a lens) of the event horizon and the darkness that lies beyond it.

However, the EHT team are not yet ready to say whether they have an equivalent image of Sagittarius A*. Part of the problem, according to Doeleman, is the real size of the black hole, which is 44 million km (27.3 million miles) across, whereas M87's black hole is 38 billion km (23.6 billion miles) in diameter. Because it is so huge, gas moving at significant fractions of the speed of light takes several days to orbit around M87's black hole, meaning fluctuations in the density and therefore light coming from that orbiting gas changes on a timescale of days. However, for the smaller Sagittarius A*, the material orbits in a matter of hours, resulting in fluctuations also on a timescale of hours, which, says Doeleman, makes calibrating the data more difficult. Add to that the fact that Sagittarius A* is not very active at all, whereas M87's black hole is enormously active, producing a jet of particles 5,000 light years long, and that makes the observations more difficult, but not impossible, and the EHT team continues to work on the problem.

Guardians of the galaxy

So many telescopes have participated in growing our knowledge of our home galaxy that it is impossible to list every one of them in detail in these pages. However, there are plenty not mentioned yet that are deserving of recognition.

The Chandra X-ray Observatory is NASA's premier X-ray telescope. Launched in 1999 as one of NASA's original quartet of 'great observatories', it observes high-energy events from exotic objects such as neutron stars, galaxy clusters and, yes, black holes. In 2018 astronomers used Chandra to detect dozens of stellar mass black holes gathering around Sagittarius A*, and theory suggests that there could be thousands more that remain undiscovered. Some of these black holes are produced by massive stars that form in clusters like the Arches or Quintuplet clusters and then explode as supernovae after a few million years, while other black holes are drawn to the galactic centre from further out by the gravitational pull of Sagittarius A*. The closer they move to Sagittarius A*, the more likely they are to run into another star and form a binary system with it. This often results in gas from the star leaking onto the black hole, which causes X-ray flares that Chandra is able to detect. Eventually, they might all end up spiralling into and merging with Sagittarius A*.

Another of NASA's original 'great observatories' is the Spitzer Space Telescope, named for astrophysicist Lyman Spitzer, who first recognised the potential for having telescopes in orbit in 1946. Operating in infrared light, it launched in 2003 and was finally shut down in January 2020 when funding ran out. Nevertheless, it left a tremendously important scientific legacy. Its infrared images of the plane of the Milky Way, and looking towards the galactic centre, are perhaps among its finest visual accomplishments, showing the dense knots and lanes of interstellar dust that weave through the galaxy's spiral arms. But Spitzer has done much more than that. It has peered into the Milky Way's star-forming nebulae and seen the youngest stars possible; it has analysed the atmospheres of exoplanets; it has mapped the Milky Way's bar following over 400 hours of careful observations; it has surveyed the star-forming zones around Gould's Belt; and it has discovered the Double Helix Nebula in the galactic centre, which appears to be two entangled filaments of gas, evidence that they have been woven by the magnetic fields that permeate the environment around Sagittarius A*.

RIGHT This infrared image reveals the many stars at the core of our galaxy and through the infrared viewpoint of the Spitzer Space Telescope. Ordinarily this region is invisible to the optical eye due to the obscuring dust. *(NASA/JPL-Caltech/S. Stolovy (SSC/Caltech))*

BELOW These 'yellowballs' were proven to be a phase in the formation of massive stars, according to data from Spitzer. *(NASA/JPL-Caltech)*

Meanwhile, NASA's Fermi Space Telescope has already been immortalised in the annals of Milky Way history by discovering the eponymous Fermi bubbles emanating from the galactic centre. Launched in 2008, gamma-ray telescopes (and, for that matter, X-ray telescopes) work a little differently to optical, infrared and radio telescopes. The latter feature large mirrors and lenses (or in the case of radio telescopes, parabolic dishes) that focus light, but gamma rays and X-rays are so energetic that they cannot be reflected or refracted in the same manner. Fermi's main instrument is the Large Area Telescope, or LAT, and it contains

18 layers of tungsten foil that interacts with the incoming gamma rays, the reaction with the tungsten atoms within the foil causing the emission of two oppositely charged particles: an electron and a positron. These particles then pass through 16 silicon-based detectors that track their motion (which mirrors the direction from which the gamma ray came) and finally strike one of 16 caesium-iodide calorimeters, which measure how much energy the particles have on impact, which in turn is directly correlated to the amount of energy the original gamma ray had.

At the opposite end to the spectrum than gamma rays are microwaves. The European Space Agency's Planck mission, which launched in 2009, was designed to observe the cosmic microwave background (CMB) radiation – the thermal afterglow of the Big Bang, as we discussed in Chapter 2. However, when Planck looked at the sky, the Milky Way was in the way. So before it could image the CMB, scientists had to subtract the microwave and far-infrared emission from the Milky Way first. But this data wasn't discarded. Instead, it was put to good use by astronomers. For example, Planck had mapped cold dust in the interstellar medium and synchrotron emission from

ABOVE The Fermi Gamma-ray Space Telescope has been operational since 2008, providing a whole new view of our galaxy and beyond. The satellite discovered the pulsar. *(NASA/Goddard Space Flight Center)*

BELOW An all-sky view of the Milky Way and how it appears at energies greater than a billion electron volts. The map is a five-year view from the Fermi telescope. *(NASA/DOE/Fermi LAT Collaboration)*

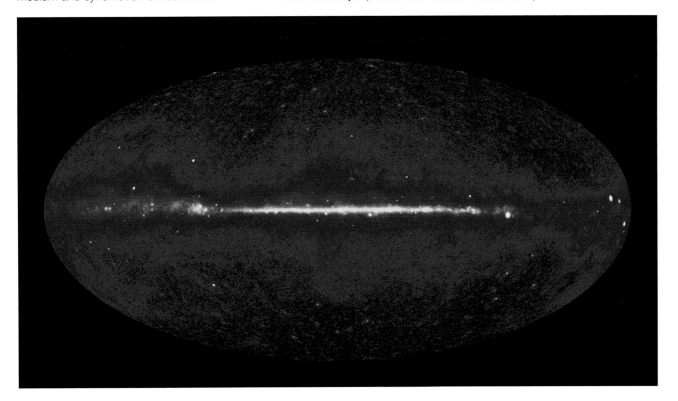

galactic magnetic fields. In one particular study, astronomers led by Cristina Popescu of the University of Central Lancashire used Planck to map the distribution of energy contained within photons of starlight in the Milky Way. Previous attempts to do this by literally doing star counts only accounted for visible and ultraviolet light, but not infrared light emitted mostly by the least massive stars, the red dwarfs, and also thermal radiation emitted by interstellar dust that is warmed by starlight.

Still to come are several next generation telescopes that are under construction and which promise to usher in a new revolution in astronomers' ability to probe the Milky Way.

When the Large Synoptic Survey Telescope (LSST) comes online it will take over as the largest, most powerful survey telescope at our disposal. It will comprehensively map the southern sky as visible from Chile. In case you are wondering, Chile – or more specifically the Atacama Desert – is a popular location for telescopes because of its low humidity, almost constantly clear sky, remoteness from centres of radio interference and light pollution, and high altitude. The LSST will make hundreds of observations of each area in the sky that it can see in its field of view, and each observation will map that area with ten times the detail of previous surveys. Then, the images of each

THE SLOAN DIGITAL SKY SURVEY (SDSS)

The most venerable of modern all sky surveys is the Sloan Digital Sky Survey (SDSS). Working from a 2.5m (8.2ft) telescope at Apache Point in New Mexico, USA, the SDSS has been running since 2000, and its principal programme for studying the Milky Way is the Apache Point Observatory Galaxy Evolution Experiment (APOGEE). It performs what SDSS scientists refer to as 'chemical cartography', whereby it spectroscopically measures the chemical abundances of hundreds of thousands of stars in the galaxy. And because it is a near-infrared telescope, it can penetrate through some of the interstellar dust in the Milky Way's thin disc to observe stars deep into the plane of the galaxy.

Some of the SDSS's most famous discoveries have involved the detection of the relics of galaxies that have interacted with and been cannibalised by the Milky Way. These obliterated dwarf galaxies have left trails of stars in their wake, which criss-cross the Milky Way and, by tracing the interloping dwarf galaxies back along these trails, astronomers are able to figure out the history of small

LEFT The dome for the Sloan Digital Sky Survey's photometric telescope. The 2.5m (8ft 2in) survey instrument sits in the background. *(Sloan Digital Sky Survey)*

area will be stitched together to build a huge mosaic, a map of the Milky Way as seen from Earth that catalogues billions of stars, their brightnesses and their colours – and hence their temperatures, which can tell us about their age and their evolution.

Being 'the biggest' is a unifying theme with all of the next major telescopes to come online in the 2020s. The European Southern Observatory's Extremely Large Telescope (ELT), with its 39m (128ft) main mirror, which is also being built in Chile on a different peak, the 3,046m (9,993ft) Cerro Armazones, and the Thirty Meter Telescope (TMT; 30m/98ft) is being built on top of Mauna Kea in Hawaii.

Each of these telescopes is by far larger than any other optical telescopes that currently exist (the largest of which are currently the 10m (32ft 10in) Keck Telescopes in Hawaii, and the 10.4m (34ft 1in) Gran Telescopio Canarias in the Canary Islands.

The ELT will gather 26 times more light than each of the 8.2m (26ft 11in) telescopes of the VLT, and will analyse the atmospheres of, and perhaps even image, extrasolar planets around neighbouring stars, as well as probe regions of star formation, supernovae remnants and star clusters in greater depth, analysing their chemical compositions to greater accuracy than ever before. On the board for the TMT's science

mergers with the Milky Way. In particular, one patch of sky was nicknamed the 'Field of Streams' in 2006 when several of these stellar streams were found to cross it. The shape of the streams has also been able to inform

scientists about the distribution and density of dark matter in the halo, with researchers concluding that dark matter in the galaxy's inner halo is distributed in a large, uniform, spherical 'blob'.

LEFT The Sloan Digital Sky Survey's 2.5m (8ft 2in) telescope. *(Sloan Digital Sky Survey)*

programme is assessing what fraction of stars in the Milky Way are binaries, triples, quadruples or even higher multiples, by resolving those systems into individual stars for which the separation is too tight for other telescopes to resolve. It will also study the development of young massive star clusters during different stages of their existence, and compare this

with how we think the globular clusters around the Milky Way formed. It will also track down low-mass red dwarf stars and white dwarfs to better map the stellar structure and chemical composition of the Milky Way's halo.

Then there's the Square Kilometre Array (SKA), which is a truly gargantuan radio telescope array being built in both South Africa and neighbouring countries, and Australia. By the time the full array is completely built, by around 2030, it will feature thousands of radio dishes for high-frequency observations and up to a million antennae for low-frequency radio wave detection, which will conduct the most detailed radio surveys of the Milky Way in the Southern Hemisphere sky, and will study the magnetic fields that writhe through the Milky Way by detecting synchrotron radiation emitted when charged particles interact with those magnetic fields.

Finally, NASA's long-awaited James Webb Space Telescope (JWST) should be launching in 2021. As a primarily infrared telescope it is not a direct replacement for the Hubble Space Telescope, but its giant (for a space telescope), segmented, 6.4m (20ft 11in) mirror will provide

astounding views of the cosmos and study with greater clarity than ever before the star- and planet-forming regions that inhabit the Milky Way.

ABOVE Ernie Wright, a NASA engineer working in the James Webb Space Telescope surveys six primary mirror segments. *(NASA/MSFC/David Higginbotham)*

LEFT The James Webb Space Telescope at the NASA Goddard cleanroom. *(NASA/Maggie Masetti)*

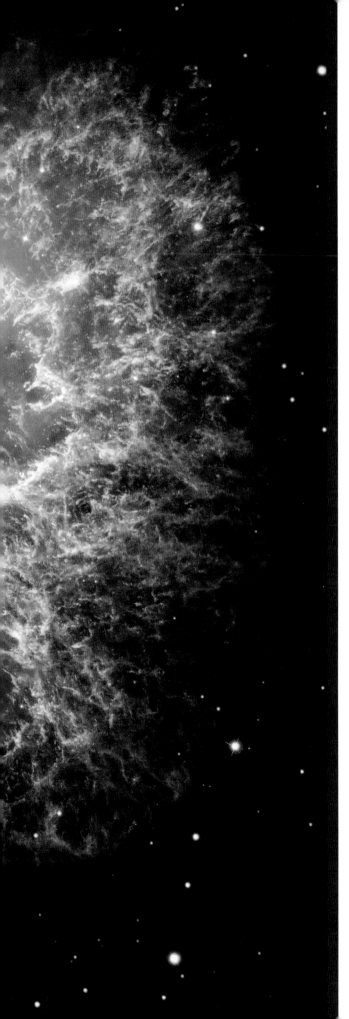

Chapter Six

Denizens of the Milky Way

You don't need a fancy telescope or a spacecraft to see the Milky Way. On a warm summer's night, best in late July or August, go outside into your back garden and look upwards. If you're lucky enough to live in an area that doesn't have much light pollution, you should see a shimmering river of faint starlight arcing overhead from the south. That is our magnificent galaxy.

OPPOSITE The Crab Nebula is the aftermath of a stellar explosion; the six-light-year-wide expanding dust has been drifting into space for nearly 1,000 years.
(NASA, ESA, J. Hester and A. Loll (Arizona State University))

We see it as a band across the sky because that is the perspective afforded us from our position within the Milky Way's disc. We see it edge-on from inside the disc. Just imagine what the view must be like from a globular cluster high above the plane of the disc in the halo!

We can also see a portion of the Milky Way's band in the winter too, but summer is the best time to view it, and not only because the weather can be pleasant. During the Northern Hemisphere's summer we are looking towards the bustling centre of the Milky Way. As the name suggests, Sagittarius A*, at the heart of the Milky Way, is in the direction of the constellation of Sagittarius, the Archer. Of course, we can't actually see Sagittarius A* from our back gardens, but we can see the dense star clouds and dark dust lanes that fill the spiral arms, as well as the colourful nebulae that are sprinkled across the spiral arms like pixie dust.

Galactic coordinates

Astronomers use two methods to map what's out there in the Milky Way. The first is an Earth-based coordinate system, which pictures the universe as sitting on a 'celestial sphere', with longitude and latitude on the celestial sphere reflecting longitude and latitude on Earth. On the celestial sphere, longitude is called right ascension, and is measured in hours, minutes and seconds (these are tangentially related to time – an hour of right ascension is the distance an object moves at its highest point in the sky as a result of Earth rotating), while latitude is referred to as declination, which is measured in degrees, minutes and seconds (here referring to arcminutes, of which there are 60 in one degree) and arcseconds (there are 3,600 in one degree).

However, this doesn't give us a true three-dimensional view. To do that, we need a system of galactic coordinates, with a galactic longitude and a galactic latitude aligned with the plane of the Milky Way's spiral disc. There are even north and south galactic poles, which are directly above and below the plane of the Milky Way.

That said, the system is still centred on the solar system, where the values of galactic longitude and latitude are zero (similarly to where the X and Y coordinates cross on a graph). Zero degrees galactic longitude is in the direction of the centre of the galaxy, in the constellation of Sagittarius (the Archer), where the black hole Sagittarius A* is. A right-angle turn (if we were taking a bird's eye view of the Milky Way) 90 degrees galactic latitude points towards Cygnus, the Swan; 180 degrees galactic latitude is in the constellation of Auriga (the Charioteer), opposite the galactic centre; and 270 degrees points towards the constellation of Vela (the Sails). Galactic latitude then measures how high above the plane of the Milky Way – as seen from Earth – an object is, with the galactic poles having latitudes of plus and minus 90 degrees.

It's fair to say that astronomers usually use the coordinates on the celestial sphere more often than galactic coordinates. That's because we can measure an object's position in right ascension and declination with 100 per cent accuracy, whereas measuring real distances and positions of objects in the universe is mired with difficulty. We've already come across the concept of standard candles, but it's only possible to use them if there is a handy one available, however many nebulae and star clusters in the Milky Way don't have them. Trigonometric parallax, as we'll describe fully later in this chapter, can provide an alternative means of measuring distance, but it is only accurate for relatively nearby objects, where the parallax angle can be resolved. Beyond that, measuring distances largely involves educated guesswork and comparisons with nearby similar objects. As such, we just don't know the galactic latitudes and longitudes of most objects in the Milky Way very well.

There are very many diverse and fascinating objects in our galaxy that are worthy of our study. So, just what are the stand-out objects that call our Milky Way home?

Cradles of young stars

Let's start in Sagittarius, because it plays host to two of the most stunning nebulae in the night sky: the Lagoon Nebula and the Trifid Nebula. Both are star-forming HII regions. HII refers to ionised hydrogen gas – ultraviolet light from the stars being born within them is

ionising (i.e. stripping atoms of electrons) the hydrogen gas in the nebula's clouds. One thing that we will quickly learn is that measurements of distances to objects such as the Lagoon and Trifid are notoriously uncertain. Absent of a standard candle to act as a milestone, estimates of their distances are, at best, educated guesses based on how much their light has been reddened by intervening dust in interstellar space. Astronomers called this reddening 'dust extinction', and the further away an object is, the more its light is dimmed and reddened. Based on this, the Lagoon Nebula, which looks like a great lake of gas, is reputedly somewhere between 4,300 and 4,900 light years from us. The distance to its neighbour in the sky, the Trifid Nebula, is even more vague, with estimates based on dust extinction ranging from as close as 2,660 light years, and as far away as 9,000 light years. If the latter is correct, then it places the Trifid in one of the Milky Way's two main spiral arms, the Scutum–Centaurus Arm.

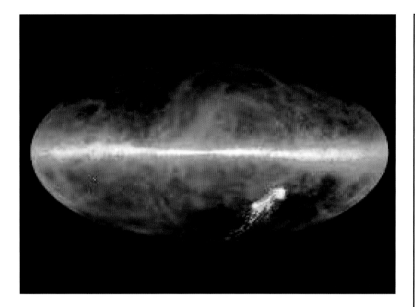

ABOVE **The entire sky visible in the light of neutral atomic hydrogen.** *(HI4PI)*

BELOW **The Coalsack Nebula.** *(Magnus Manske)*

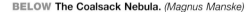

The Trifid Nebula was so-called because, when John Herschel observed it in the 19th century, he saw three distinct lobes split by a dark central dust lane, but modern telescopic technology has shown us that the dust lane really splits the nebula into four lobes. What marks the Trifid out as special is how it encapsulates the three main kinds of nebula: it's a HII emission nebula, but also a reflection nebula, scattering the light of the bright, young stars up to 20 times the mass of our Sun that emerge from the gas clouds, and it's also what astronomers refer to as a 'dark nebula', in the form of the gas embedded with thick, black dust.

A young star cluster, designated NGC 6530, is seen emerging from the Lagoon Nebula. This may very well have been how the Sun was born, 4.6 billion years ago – see Chapter four for more detail. Meanwhile, there's still plenty of material remaining in the nebula to continue forming stars across its 115 light year extent.

We'll discuss the role of dark nebulae in the Milky Way later in this chapter.

The Great Nebula

The most famous nebula, and one that is easily visible to the naked eye as a fuzzy 'star' in the sky, is the Orion Nebula, located 1,344 light years away from us. Look for it in the constellation of Orion, hanging just below the three distinctive 'Belt' stars.

With it being so close to us, the Orion Nebula has been studied in great depth and has been shown to be a cauldron of all manner of astrophysical processes. On the one hand there's the fact that it is busily forming stars, and has been doing so for the past 15 million years. Underlying the famous stars of Orion is a vast molecular cloud complex, a GMC that has spawned tens of thousands of stars. The first burst of starbirth took place in a region north-west of Orion's Belt, but back then the asterism didn't exist. This initial round of star formation

RIGHT A detailed shot of the star-forming region the Orion Nebula (Messier 42) captured by NASA/ESA's Hubble Space Telescope. *(NASA, ESA, M. Robberto (Space Telescope Sci9ence Institute/ESA) and the Hubble Space Telescope Orion Treasury Project Team)*

produced the Orion 1a group of stars. The most massive of these stars have already been and gone, exploding as supernovae and clearing out large swathes of the gas in the surrounding cloud, although lower mass stars remain. However, the supernova shocks compressed gas in the cloud towards the south-east, producing a second burst of star formation that resulted in the creation of, among others, the three Belt stars – Alnitak, Alnilam and Mintaka. This trio, and their thousands of siblings, are known as the Orion 1b group.

ABOVE A section of the famous Orion Nebula, 1,350 light years away from Earth, in infrared and millimetre wavelengths from the Very Large Telescope and the Atacama Large Millimeter/submillimeter Array. *(ESO/H. Drass/ALMA (ESO/NAOJ/NRAO)/A. Hacar)*

LEFT The Ophiuchi Cloud Complex. *(Magnus Manske)*

ABOVE This NASA Wide-field Infrared Explorer (WISE) image of the Rho Ophiuchi Cloud Complex reveals a wealth of astronomical objects. *(NASA/JPL-Caltech/WISE Team)*

BELOW A complex of molecular clouds and star-forming regions, located over 6,000 light years away. The collection of clouds make up three regions known as W3, W4 and W5. *(ESA/Herschel/NASA/JPL-Caltech, CC BY-SA 3.0 IGO; Acknowledgement: R. Hurt (JPL-Caltech))*

Again, the most massive of the stars in Orion 1b exploded, compressing the surrounding molecular gas in two new areas. One was eastwards, towards a part of the GMC called the Orion B cloud, which gave birth to another group of stars, including Sigma Orionis, about 2–4 million years ago. Sigma Orionis is the hottest and most luminous of this group of stars, its radiation illuminating the western rim of the Orion B cloud as the star's ultraviolet light gradually eats into the cloud, eroding it. Meanwhile, to the south, star formation ignited in another part of the old GMC, called the Orion A cloud, about five million years ago. This included the dense blister of star formation that is the Orion Nebula, resulting in another generation of stars referred to as the Orion 1c group.

As with previous episodes of star formation in the Orion GMC, eventually the Orion Nebula will fizzle out, its star-forming material exhausted and its remnants blown away by the combined radiation of the stars it has bequeathed and the shocks of some of their supernovae. However, this will not be the end of star formation in the constellation. The Orion A cloud is 40 light years long, and the Orion Nebula makes up only half of that. The shocks from the destruction of the stars we currently see shining in the Orion Nebula, particularly those belonging to the Trapezium Cluster at the nebula's heart, could compress gas further down the cloud, so that as the Orion Nebula fades, a new star-forming region will arise. The king is dead, long live the king!

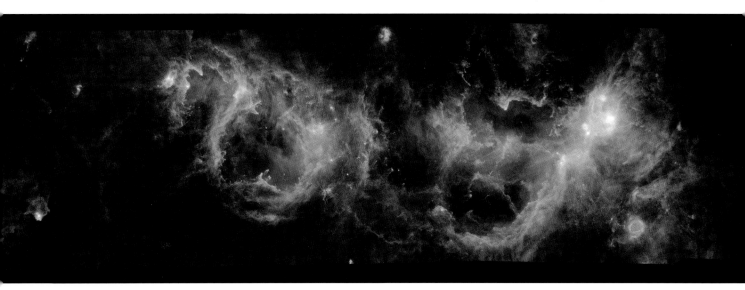

Further afield, the stellar radiation winds and supernova shocks from the stars formed in Orion over the past 15 million years have blown a gigantic superbubble of hot, ionised gas. Although this superbubble is directly visible in X-ray light, as it expands it is ploughing into the cooler gas of the interstellar medium, sweeping it up into a curving arc of nebulosity wrapped around part of the bubble, which astronomers call Barnard's Loop after the 19th century astronomer, Edward Emerson Barnard (1857–1923).

Right now, the main star-forming action is taking place in the Orion Nebula, and the direct consequence of that is the impressive Trapezium star cluster. Less than 100,000 years old, the Trapezium's name comes from the pattern in the sky of its four brightest members, which weigh in at between 15 and 30 solar masses. They are joined by about 2,000 other newly born stars distributed across about 20 light years. The four brightest stars are more compact, spaced within a volume just 1.5 light years across.

But peculiar things can happen to stars that are in such close quarters in young clusters. Let's momentarily head into two completely different constellations, Auriga, the Charioteer and Columba, the Dove. Here we encounter two stars, named AE Aurigae and Mu Columbae, that have extraordinary motions through space, travelling at 137 and 141 km/s respectively. Tracking their trajectories backwards through space, we astonishingly find that both hail from the Orion Nebula, despite now being hundreds of light years away. Even more astonishingly, the two stars come from the same cataclysmic event that produced the binary – i.e. double – star system Iota Orionis, just to the south of the nebula. Based on the velocities of

the 'runaway' stars, AE Aurigae and Mu Columbae began their journey from the Orion Nebula 2.6 million years ago. So what happened?

It seems that once upon a time, each star – AE Aurigae and Mu Columbae – was in its own binary star system that existed within the cramped confines of their birthing cluster, which is called NGC 1980. The two double systems ended up getting too close to one another, and in the gravitational dance that ensued, AE Aurigae and Mu Columbae were viciously flung away from their siblings. Meanwhile, their remaining companions joined forces to form the Iota Orionis system. Further evidence for this comes from the fact that we know that the

ABOVE A runaway star creates a bow shock as it plunges through the interstellar medium of the Milky Way.
(NASA/JPL-Caltech)

two stars of Iota Orionis did not form together, because one of them is older than the other, and the respective ages match those of AE Aurigae and Mu Columbae.

There are other high-speed runaways coming from the various star-forming zones across Orion. Even Betelgeuse, which is the bright, volatile, red supergiant star in the upper left corner of Orion's star pattern, originated near the Orion 1a star group 10 million years ago. Betelgeuse will explode as a supernova sometime in the next million years. Given that the Orion star-forming zone is nothing special, and that there are, and have been in the past, many similar star-forming regions, then it implies that the Milky Way is full of these refugee stars forced out of their star clusters by the gravitational interactions with neighbouring star systems.

Astronomers have been able to detect planets forming within the Orion Nebula

ABOVE If stars are able to gain enough speed, they can escape the gravitational clutches of a galaxy. They're known as hypervelocity stars.
(NASA, ESA and G. Bacon (STScI))

BELOW The Pistol Star sits close to the core of our galaxy.
(Don F. Figer (UCLA) and NASA)

AN UNORDERED SOLAR SYSTEM

It's not just stars that are unceremoniously shown the door. Smaller objects such as planets, comets and asteroids are under even greater threat from these kinds of interactions. Astronomers have identified free-floating planets in the vicinity of the Trapezium, rogue worlds destined to wander the darkness of interstellar space forever. In 2017, astronomers detected a small asteroid or comet (the jury is still out on which one it was) that came from interstellar space passing through our solar system. Named 1I/2017 U1, or 'Oumuamua, it is predicted to be one of 10,000 interstellar objects passing through our solar system at any one time. All of these objects have been liberated from their home systems by some force, be it passing stars, close encounters with other large planets, or even supernova explosions that blow them away rather than vaporising them. But to cut a long story short, essentially the Milky Way is littered with waifs and strays of all sizes, from full-blown stars to tiny asteroids. It's a messy environment that hints at the chaos that ensues among the denizens of the galaxy.

before they get the chance to go rogue. Closer examination of the nebula by the Hubble Space Telescope has revealed objects known as 'proplyds', which are discs of gas and dust wrapped around very young stars that are still in the process of forming. Nearly 200 proplyds have been identified in the Orion Nebula, and it is speculated that these discs, which occur naturally when a collapsing cloud of gas that is condensing to become a star begins spinning and flattening out as a result, contain embryonic planets hidden by the dust in the disc.

The Pillars of Creation

While gas and dust in the Milky Way remain dormant inside inactive giant molecular clouds, the GMCs are just amorphous blobs. However, once they light up with starbirth, these clouds are sculpted by the ionising radiation coming off those stars. In a way, the stars are like painters, the stellar winds are their brushes and their canvases are the mighty GMCs themselves. They're like the Banksy of interstellar space.

If you need proof, then just look at the

ABOVE Betelgeuse, a red supergiant in the constellation of Orion (the Hunter) is shown in millimetre wavelengths in this view of its surface. *(ESO/NAOJ/NRAO)/E. O'Gorman/P. Kervella)*

BELOW Star-forming regions in the southern Milky Way, NGC 3603 (left) and NGC 3576. *(ESO/G.Beccari)*

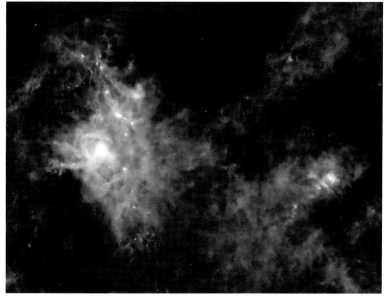

Eagle Nebula, which is a huge region of star formation, far larger than the Orion Nebula, in the constellation of Serpens (the Serpent). It's 7,000 light years away and famous for the Hubble Space Telescope image of three columns of dense gas and dust that exist within the heart of the nebula, collectively known as the Pillars of Creation. Each 'pillar' is several light years long (the longest extends four light years, so that it would span the distance between the Sun and Alpha Centauri, which is the closest star system to the solar system). They look remarkably like the towering wind-eroded rocks that stand above desert sands on Earth, only in the Pillars' case, the eroding wind is ultraviolet light from newborn stars.

ABOVE An infrared view of the Aquila Rift, located at a distance of about 850 light years away and which forms part of the Gould Belt; a giant ring of stars and star-forming clouds in the vicinity of the Sun. *(ESA/Herschel/SPIRE/PACS/Ph. Andre (CEA Saclay) for the Gould Belt survey Key Programme Consortium)*

BELOW A sequence of star-forming regions within the molecular cloud W48 in the constellation of Aquila (the Eagle), which forms part of the Gould Belt. *(ESA/Herschel/PACS/SPIRE/ HOBYS Key Programme Consortium)*

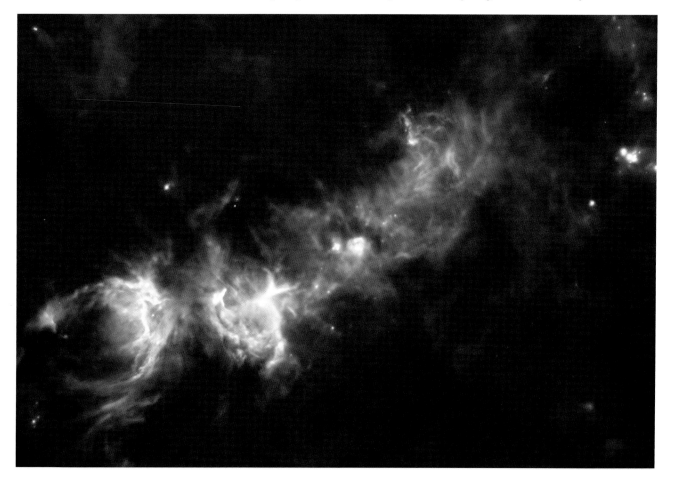

RIGHT The stellar nursery G305 shines brightly in infrared light. *(ESA/Herschel/PACS, SPIRE/Hi-GAL Project. Acknowledgement: UNIMAP / L. Piazzo, La Sapienza – Università di Roma; E. Schisano / G. Li Causi, IAPS/INAF, Italy)*

Inside each of the Pillars, stars are forming. The dust in the molecular gas helps insulate the gas, keeping it cold enough – no warmer than 10 degrees above absolute zero – for pockets of the gas to begin to gravitationally collapse to form new stars. Look closely at the head of each Pillar and 'knobbly' protrusions on stalks can be seen. Each protrusion, called an evaporating gaseous globule (or EGG for short), is about the size of the solar system, and indeed is a cocoon for a brand new star system in the making. Because the EGGs are formed of dense gas and dust, it takes longer for the ultraviolet radiation to ionise the atoms in the gas molecules, and hence they do not erode as quickly. Eventually, though, the EGGs too will evaporate as the burgeoning star inside them lights up.

Indeed, the Pillars themselves are not forever. The erosion by stellar radiation will eventually whittle them down, and should a supernova explode in their vicinity, which is eminently possible, then the resulting shockwave could destroy the Pillars, shattering them to pieces. Indeed, Nicolas Flagey of the Institut d'Astrophysique Spatiale in France has proposed that a supernova has already exploded nearby, based on the existence of an anomalous cloud of hot gas and dust near the Pillars. Flagey suspects this is the expanding cloud of debris from the supernova, and that in a thousand years' time it will have destroyed the feature. However, other astronomers are sceptical, pointing out that a supernova remnant should leave more evidence in the form of X-rays and radio waves. If so, the Pillars may still be around for a long time to come.

RIGHT The Christmas Tree Cluster and Cone Nebula (NGC 2264) through the Wide-Field Imager at La Silla Observatory, some 2,400m (7,874ft) high in the Atacama Desert, Chile. *(ESO)*

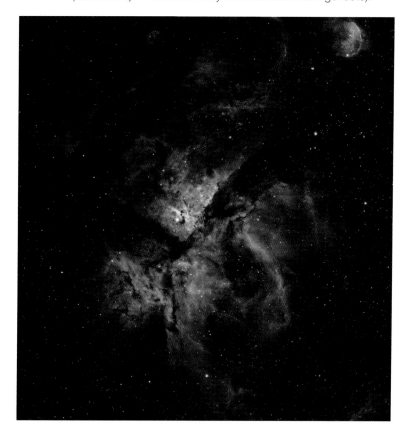

Eta Carinae

Another impressive star-forming HII region is the Carina Nebula, which exists 7,500 light years away in the constellation of Carina, the Keel (of the mythical ship, the Argo, as from the story of Jason and the Argonauts).

One of the largest star-forming regions in the entire Milky Way, the Carina Nebula is over 300 light years across. Hubble Space Telescope images of the nebula just show an immense wall of molecular gas, dotted with numerous young star clusters and nebulous subcomponents, such as the Keyhole Nebula, which is a cloud of gas shaped like a keyhole. A massive V-shaped dark nebula, formed of dust that blocks starlight, appears to split the Carina Nebula in two, but it's really just part of the same vast, gaseous complex.

What the nebula is most famous for, however, is a star system that is of an exceedingly rare type. The star Eta Carinae is actually a binary system, with one of its member stars possessing over 100 solar masses, easily one of the most massive – if not the most massive – star in the Milky Way (but still only a third as massive as R136a1 in the Large Magellanic Cloud). Meanwhile, its companion star is hefty in its own right, with a mass about 30 times greater than the Sun.

But Eta Carinae's real claim to fame came in 1843, when it was seen to mysteriously brighten to temporarily become the second brightest star in the entire sky – only Sirius shone more brilliantly. As it grew more luminous, it spat out up to 40 solar masses worth of material to form an hourglass-shaped nebula around it, called the Homunculus Nebula. Despite the violence of this outburst, the star

had not exploded – it's still there today – so what happened? Nobody is entirely sure. Some astronomers suspect that once upon a time there may have been three stars in the Eta Carinae system, but that two of them crashed and merged into one another, forming the giant 100-solar-mass primary star and producing the outburst. A credible alternative is that Eta Carinae is what is known as a 'supernova imposter'. Some stars in other galaxies have been seen to explode in a supernova, only to somehow miraculously survive because they then explode again, for real, a few years later. However, as we've stated, Eta Carinae is still

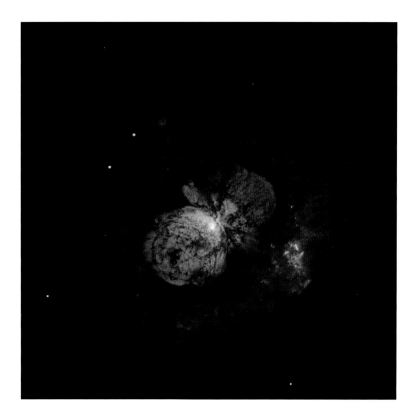

there, whereas other supernova imposters have only taken a few years to go supernova for real.

Either way, Eta Carinae definitely will go supernova for real eventually, but despite its extreme mass, there's no guarantee that it will put on a light show when it does. Some massive stars can collapse directly to become a black hole, releasing no light at all because the black hole swallows it all. Others explode in what is called a pair-instability supernova, in which pairs of electrons and positrons are produced inside the star when atomic nuclei and gamma rays released by fusion processes in the star's core collide. This leads to a drop in radiation pressure holding the star up against gravity as the gamma-ray photons are absorbed or lose energy during the collisions. The star explodes in a supernova even more luminous than normal, leaving behind a black hole produced by the star's collapsed core. It's even possible that when Eta Carinae explodes, it will result in a gamma-ray burst (GRB), wherein two jets of powerful gamma rays are emitted along the star's magnetic axis.

GRBs are so luminous in gamma rays that they are seen across the universe, and even their visible light fireballs can be bright enough to be seen across billions of light years. The main type of GRB is known as a long-duration GRB, which can spend several minutes emitting powerful gamma rays, and are the result of exploding stars. Scientists trying to understand the triggers for a GRB, as opposed to a normal supernova, have noted that galaxies with low metallicities – that is, a low abundance of heavy elements – tend to harbour long-duration GRBs rather than more chemically evolved galaxies like the Milky Way.

While on the face of it that is good news for life in the Milky Way, since the radiation from GRBs is deadly enough to sterilise whole planetary systems, it doesn't mean we're out of the woods. Star-forming dwarf galaxies like the LMC are prime locations for harbouring GRBs. There's also another type of GRB, called

BELOW A two-light-year long nebula located within the Carina Nebula, named the 'Finger of God'. *(NASA, ESA, N. Smith (University of California, Berkeley) and the Hubble Heritage Team (STScI/AURA); credit for CTIO Image: N. Smith (University of California, Berkeley) and NOAO/AURA/NSF)*

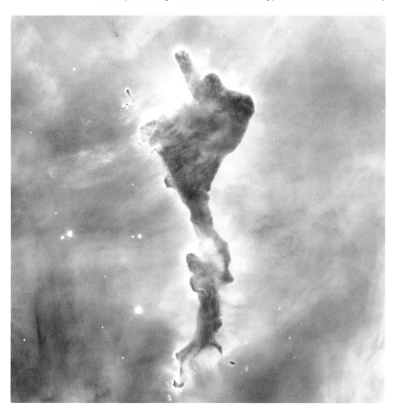

RIGHT Cygnus X-1 is a binary system, comprising a black hole with a mass of around ten times the Sun that's siphoning a blue supergiant. The result can be seen in X-rays in this image from NASA's Chandra X-ray Observatory. *(NASA/CXC)*

short-duration bursts, which are produced when two neutron stars collide. These cosmic blasts last for less than two seconds, but still pack a powerful punch. Neutron stars are produced when massive stars explode as a normal supernova, but aren't quite extreme enough for their core to collapse to become a black hole. Instead, their core collapses down to a size of between 10 and 30km, and dense enough that the electrostatic barrier between the positive charge on protons in atomic nuclei and the negative charge on orbiting electrons is overcome, and protons merge with the electrons, the charges cancelling out to form neutral neutron particles – hence we have a neutron star.

Because many stars, especially massive stars, like to form in binary pairs, they often leave behind binary systems of orbiting neutron stars. Eventually their orbits decay and they

BELOW An artist's impression of the binary system of recurrent nova RS Ophiuchi. It depicts a short time after the white dwarf (right) has exploded as a nova – the other star is a red giant. *(Casey Reed)*

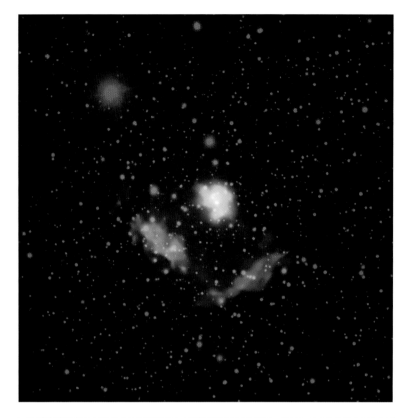

LEFT The youngest known pulsar in the Milky Way, Kes 75. *(X-ray: NASA/CXC/NCSU/S. Reynolds; Optical: PanSTARRS)*

spiral into a collision, and hey presto, a short GRB is released. It's not just gamma rays either; gravitational waves have been detected coming from neutron star collisions, while the fireball they produce is called a kilonova.

Although all short GRBs detected so far have been hundreds of millions, if not billions of light years away, eventually one will occur in our galaxy.

The Crab Nebula

There are plenty of neutron stars in the Milky Way, and even some binary neutron stars, such as the famed binary pulsar, a.k.a. PSR J0737-3039, which is between 3,000 and 5,000 light years distant. A pulsar is just a spinning neutron star that emits beams of radiation along their magnetic axes. As the pulsars rotate, these beams spin around, giving the impression of pulses as they periodically flash in our direction. Most pulsars spin once every second or two, but some can spin hundreds of times per second. They were discovered in 1967 by Jocelyn Bell, who was a PhD student in Cambridge at the time. Her radio astronomy experiment detected strange, regular signals, essentially going beep-beep-beep. Bell's PhD supervisor, Antony Hewish, realised that these strange signals had been predicted as evidence for pulsars, but until Jocelyn Bell and her experiment, nobody had ever detected one. As it turns out, although they are powerful at radio wavelengths, pulsars emit across the electromagnetic spectrum.

Eventually, pulsars slow down their spin, switch their beams off and become normal

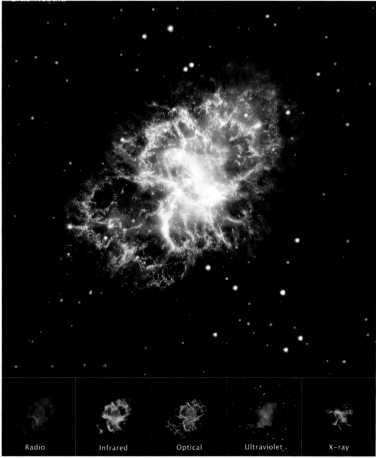

Radio   Infrared   Optical   Ultraviolet   X-ray

LEFT The Crab Nebula through different wavelengths: radio, infrared, optical, ultraviolet and X-ray. *(NASA, ESA, G. Dubner (IAFE, CONICET-University of Buenos Aires) et al.; A. Loll et al.; T. Temim et al.; F. Seward et al.; VLA/NRAO/AUI/NSF; Chandra/CXC; Spitzer/JPL-Caltech; XMM-Newton/ESA; and Hubble/STScI)*

RIGHT This column of cool molecular hydrogen gas and dust births new stars inside the iconic Eagle Nebula (Messier 16). *(NASA)*

neutron stars. As such, we're more likely to find pulsars in the midst of supernova remnants, and one of the most well-known is the Crab Pulsar, which is found inside the Crab Nebula in the constellation of Taurus, the Bull. As you might suspect, the Crab Nebula isn't really a nebula like the Orion or Eagle nebulae; rather, it is the debris from a nine–11-solar-mass star that exploded as a supernova in the year AD1054, a brilliantly explosive event that was actually seen by Chinese astronomers. It shone in the daylight sky for 23 days, reaching magnitude -6 at its peak, which is brighter than the planet Venus, and was visible to ancient astronomers in the night sky for over 21 months before it faded away.

It's astonishing to think that the supernova could be seen in the daylight, yet it was located 6,500 light years away in the Milky Way's Perseus spiral arm, the Crab Nebula remnant that it left behind has grown to be 11 light years across, and is still expanding at a velocity of 1,500km (932 miles) per second. Its name comes from the observations of the supernova remnant in the 1840s by Irish astronomer William Parsons, who was the third Earl of Rosse and who had a giant 72in (1.8m) aperture telescope called the Leviathan, and thought that the nebula was shaped a little bit like a crab as seen through his telescope.

The Crab Nebula can be split into two zones. The inner zone glows blue as a result of synchrotron radiation, wherein electrons accelerating away in the pulsar wind, reaching half the speed of light, find themselves curving around the tense magnetic field lines emanating from the pulsar itself, and as they move around the field lines they emit the synchrotron radiation, which glows at high energies, in blue light, ultraviolet, X-rays and gamma rays.

The outer zone, meanwhile, glows red and

RIGHT The Pillars of Creation inside the Eagle Nebula, also known as the Eagle Nebula (Messier 16). *(NASA, ESA and the Hubble Heritage Team (STScI/AURA))*

THE MIGHTY CRAB NEBULA

The Crab Nebula is an amazing natural particle accelerator, far outstripping the energies generated at the Large Hadron Collider at CERN. The pulsar, which is seen to pulse once every 33 milliseconds as it rotates 30 times per second, has a gravitational field that is 100 billion times stronger than the gravity at the surface of the Earth (which isn't surprising, when you consider that the pulsar has up to twice the mass of the Sun packed into a sphere about the size of a large city). However, when the core of the progenitor star (i.e. the star that exploded) collapsed, it compacted the star's own magnetic field, rendering it incredibly intense and generating an accompanying electric field of ten quadrillion volts (that's 10,000,000,000,000,000 volts). This electric field is so powerful that it can even overcome the crushing gravity of the pulsar and drive away charged particles – electrons and protons – held near the surface of the pulsar. Indeed, the Hubble and Chandra space telescopes have observed concentric rings surrounding the pulsar up to a distance of a light year from it, and composed of charged particles flung out by the pulsar. However, most of the charged particles find themselves either in the pulsar beams or becoming the pulsar wind – a flux of radiation flowing out from the pulsar and slamming into the gaseous debris from the supernova explosion.

charged particles gain tremendous amounts of energy, enough that they eventually are able to escape the confines of the supernova remnant and race out into space, becoming what we call 'cosmic rays' that fill the void between the stars in the Milky Way.

The pulsar's rotation is slowly but surely slowing down, as rotational energy is removed and turned into the energy that drives the escaping charged particles and the synchrotron radiation. Eventually, it will rotate so slowly that its beams will lose power and the pulsar will transform into just a regular neutron star. However, because we know the age of the pulsar and the surrounding remnant, almost to the day, we're able to put a time-stamp on the behaviour of the pulsar. There are theoretical models about how slowly pulsars decrease their spin rate with age, and how that relates to their luminosity and the strength of their magnetic field. The Crab Pulsar then becomes the standard by which other pulsars in the Milky Way can be compared. There are approximately 2,000 that have been discovered, all in either the Milky Way or the nearby Magellanic Clouds, except for one that has been discovered in the Andromeda Galaxy. However, it is expected that there are a billion neutron stars lurking somewhere within our galaxy, including the Geminga neutron star that is one of the closest to us at a distance of 800 light years, so there must be many more still in the pulsar stage waiting to be discovered.

Similarly, the Crab Nebula also acts as a standard by which other supernovae remnants can be compared. Besides the Crab, the known supernova remnants in the Milky Way include the Veil Nebula, Cassiopeia A, which is less than

BELOW Taken by NASA's Galaxy Evolution Explorer, red giant variable Mira A sports a comet-like tail and reveals that the star is blowing off hydrogen gas. *(NASA)*

green, resulting from the expanding outer layers of the remnant crashing into the surrounding interstellar medium (ISM). The red glow is produced by hydrogen-alpha light or, in other words, emission from excited hydrogen atoms, while the green glow is the result of emission from doubly-ionised oxygen atoms (OIII). The pulsar wind follows, and when the charged particles reach the shock front, they begin bouncing around it, following the turbulent magnetic field lines where the expanding remnant crumples as it starts to hit the denser ISM. By bouncing between the field lines, the

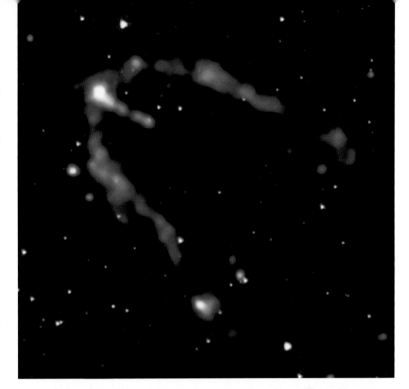

400 years old, and remnants of supernovae witnessed by Tycho Brahe in 1572, and then by his former assistant, Johannes Kepler, in 1604.

The Ring Nebula

Not all stars die as supernovae. If a star has less than eight times the mass of the Sun, its eventual expiry is a far gentler, and even prettier, affair, and this will be the fate that will also befall the Sun.

When a Sun-like star runs out of hydrogen deep within its core, the core begins to shrink, collapsing under gravity because the star's core is not able to produce as much energy as before, and it is that energy, radiated outward from the core and providing an upwards pressure on the layers of the star around the core, that is holding the star up. As the core then shrinks, the density and therefore the pressure and the temperature within the core also rise, great enough that the star can begin fusing helium instead, in an event known as the helium flash, to produce oxygen and carbon atoms, while the temperature increases in the star's outer layers to become hot enough for

RIGHT The supernova remnant Cassiopeia A in false colour, featuring observations from the Hubble Space Telescope, Spitzer and the Chandra X-ray Observatory. (Courtesy NASA/JPL-Caltech)

LEFT The Veil Nebula, also known as NGC 6960, glows with the emission of hydrogen, sulphur and oxygen. (Ken Crawford)

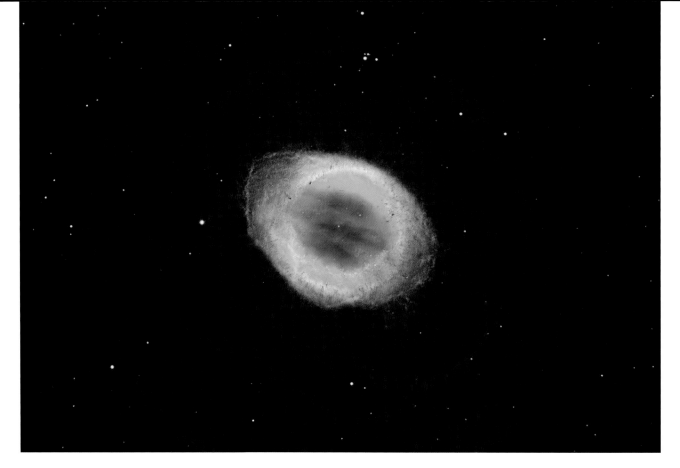

ABOVE **The Ring Nebula.** *(NASA, ESA, C. R. Robert O'Dell (Vanderbilt University), G. J. Ferland (University of Kentucky), W. J. Henney and M. Peimbert (National Autonomous University of Mexico) Credit for Large Binocular Telescope data: David Thompson (University of Arizona))*

hydrogen fusion to take place there. The energy produced is enough to cause the outer layers of the star to dramatically expand in size many times over, and as they expand they grow less, so the outer layers of the star cool. The star turns into a cool red giant, but the radiation winds from within have gusts strong enough to blow the now-diffuse outer layers from the star completely. Like a reptile shedding its skin, the outer layers drift into space to become a so-called planetary nebula. Meanwhile, the star's core grows inert, and is left exposed by the loss of the outer layers. We call such stellar cores 'white dwarfs'. They are about the size of Earth, and are exceptionally hot.

The name 'planetary nebula' can be derived to observations made by William Herschel, who compared the fuzzy appearance of planetary nebulae through his telescope as looking a bit like the discs of planets, in terms of apparent angular size (not real physical size) and uniformity of brightness. One of the most famous is the Ring Nebula, seen in the constellation of Lyra (the Harp).

Located 2,570 light years away, the Ring Nebula's appearance is deceiving. What we see is a flat, oval-shaped ring of light. However, careful observations by the Hubble Space Telescope have shown that this light-year-wide nebula is actually bipolar. The oval ring is actually a doughnut-shaped torus seen reasonably edge-on (at an angle of about 30 degrees, to be precise). The blue region inside the doughnut – the jam filling, if you will – is a rugby-ball shaped pocket of less dense gas pointed nearly head-on towards us. If we could look from a more forgiving angle, we'd see the cosmic rugby ball poking out through the hole in the doughnut.

The reason for the bizarrely complex shape is uncertain. We do know it is related to outflows from the central star in its death throes, and one popular theory is that planetary nebulae like this are actually binary systems, where the gravity of the companion star, which has not yet reached old age because it contains less mass and therefore lives longer, is able to sculpt the material coming off the dying star and into the nebula, directing where the outflows go. Given that studies have shown that a quarter to a third of all star systems are binaries, triples or even quadruples, then the possibility is not that outlandish. It's even been speculated that a single star, like the Sun, will not produce a dramatic planetary nebula as pretty as the Ring, because it doesn't have a companion star to do the sculpting for it.

What is left of the dying star at the centre of the Ring Nebula is an object transitioning to becoming a white dwarf, made mostly of carbon and oxygen according to the emission lines in its spectrum. The white dwarf has a mass of about 0.6 Earth's mass. Given that most of the star's material escaped into space, that fact that so much is still left in the form of the white dwarf means that the original star was more massive than the Sun, with up to 2.2 solar masses. White dwarfs are born hot – they've been inside a star, so little wonder – and the Ring Nebula's white dwarf has a surface temperature of 150,000°C (270,000°F).

At the nebula's perimeter, there's lots happening. We see the ring of gas as denser than the blue medium in the nebula's centre, because it is at the rim of the nebula that the gas is crashing into the interstellar medium, causing it to pile up. Despite his, it is still expanding at 30km per second, but it is interesting to compare this expansion rate with the expansion rate of the Crab Nebula, which is much faster at 1,500km (932 miles) – a clear sign of the difference in the energy released between the very different deaths of the stars that produced each nebula. Beyond the bright ring are fainter loops and arcs, visible in long-duration exposures, the remnants of the first layers of the star to have been ejected into space.

Noticeable along the inside edge of the Ring are various tadpole-shaped knots and filaments, generally a little darker than their surroundings and pointing radially away from (or towards, depending on your viewpoint) the white dwarf. These are pockets of denser gas that have yet to be whittled away by the hot wind of radiation coming off the dying white dwarf.

Such tadpole-shaped knots, which are more properly known as 'cometary knots', are common features in planetary nebulae, and are seen to an even greater extent within the Helix Nebula, which is the closest planetary nebula to us at a distance of 650 light years away in the constellation of Aquarius. The Helix has upwards of 20,000 cometary knots and, despite their name, they are not thought to be related to real comets – the comets would have to be the size of Earth to be visible, which is highly unlikely. Instead, the name evokes the presence of the tails that give the tadpole effect.

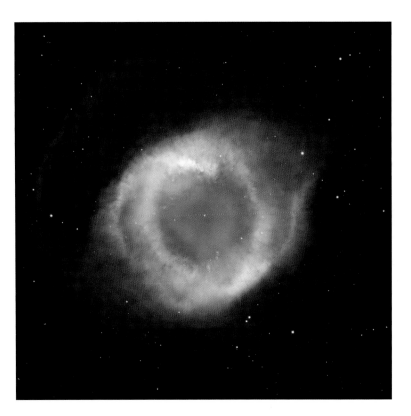

ABOVE The Helix Nebula (NGC 7293), a large planetary nebula in Aquarius.
(NASA, ESA and C.R. O'Dell (Vanderbilt University))

BELOW Zoomed-in view of the gas and dust in the Helix Nebula.
(NASA, NOAO, ESA, the Hubble Helix Nebula Team, M. Meixner (STScI); and T.A. Rector (NRAO))

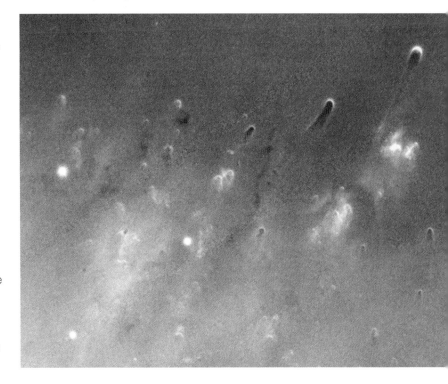

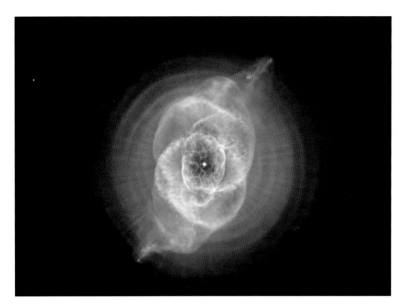

ball with a doughnut – and there are numerous other planetary nebulae that we see in the Milky Way with a similar structure, for example the Dumbbell Nebula that is 1,200 light years away in the constellation of Vulpecula (the Fox), and the even more complexly structured Cat's Eye Nebula, which is 5,300 light years away in Draco (the Dragon). The Cat's Eye Nebula features a rugby-ball shaped pocket of gas inside two conjoined spherical bubbles, surrounded by concentric rings, and two twisting tails that could be produced by a binary stars system inside, directing the outflows into two jets.

Estimates suggest that there are about 20,000 planetary nebulae in the Milky Way galaxy at any one time, although only approximately 1,800 have been discovered so far. These billowing nebulae are rich in carbon, oxygen and nitrogen in particular, and can deposit several solar masses worth of chemically-enriched gas and dust into the interstellar medium to be reborn into the next generation of stars and planets. Supernovae are even more productive suppliers of heavy elements; SN 1987A in the Large Magellanic Cloud, for example, ejected enough dust into space to build 200,000 Earth-sized planets! So don't be sad when you think of stars dying – in a kind of reincarnation, their gas and dust comes to life again in new stars and planets and, without dying stars, we wouldn't be here now.

The Helix Nebula is much older than the Ring Nebula. Whereas the latter is just 1,600 years old, the Helix is thought to have been around for 10,600 years. It has expanded such that its outermost ring is 5.74 light years in diameter, and is still expanding at 40km per second (25 miles per second). Eventually, like all planetary nebulae do, it will dissipate into the interstellar medium.

The Helix looks very similar to the Ring Nebula in terms of its structure – another rugby

Extrasolar planets

When we think of travelling through the Milky Way in a spaceship, we tend not to think of exploring the nebulae, pulsars and star clusters that we come across. Instead, science fiction focuses on the planets that all that dust put out by dead stars form. Yet until the 1990s, the only planets that astronomers knew of in the Milky Way were the ones in our own solar system.

Today, astronomers have discovered over 4,000 extrasolar planets (known for short as 'exoplanets') with that figure set to massively jump up during the 2020s, as discoveries pile up

LEFT **Star cluster Trumpler 14 is one of the hottest and most massive collections of bright stars in the Milky Way.** *(NASA & ESA, Jesús Maíz Apellániz (Centro de Astrobiología, CSIC-INTA, Spain))*

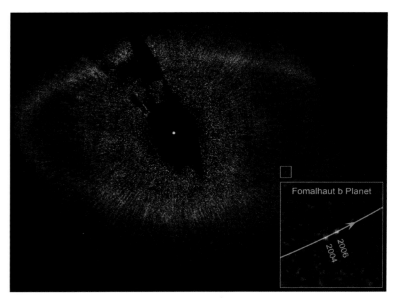

from NASA's Transiting Exoplanet Survey Satellite (TESS), the European Space Agency's PLAnetary Transits and Oscillations (PLATO) satellite and various other ground-based searches.

In Chapter four we briefly covered the general types of planets: the super-Earths, the mini-Neptunes and the hot Jupiters. Now it's time to get to know some specific examples of each type.

The closest known exoplanets to us in the Milky Way are Proxima Centauri b, which orbits the nearest star to us 4.2 light years away, and

ABOVE LEFT Artist's impression of a hot Jupiter, gaseous worlds that tightly orbit their stars. *(NASA/Ames/JPL-Caltech)*

ABOVE A coronagraph of the star Fomalhaut, its ring of dust and the location of Fomalhaut b. *(NASA, ESA, P. Kalas, J. Graham, E. Chiang, E. Kite (University of California, Berkeley), M. Clampin (NASA Goddard Space Flight Center), M. Fitzgerald (Lawrence Livermore National Laboratory), and K. Stapelfeldt and J. Krist (NASA Jet Propulsion Laboratory))*

BELOW An artist's impression of exoplanet Fomalhaut b, orbiting its star Fomalhaut. *(ESA, NASA and L. Calcada (ESO for STScI))*

IS THERE ANOTHER EARTH-TYPE PLANET OUT THERE?

We don't know yet of any planet that is a spitting image of the Earth, but several good candidates have been found. NASA's Kepler Space Telescope, which operated between 2009 and 2018, spent a good proportion of its first four years in space staring at a small patch of the Milky Way, back along the Orion Spur and into the Sagittarius Arm and containing 150,000 visible stars as it searched for the transits of exoplanets around them (the final five years were as part of its more varied K2 mission, which it was forced into when two of its reaction flywheels, used for 3-axis attitude control to maintain accurate pointing towards its original target stars, failed). These stars were all much further away than Proxima Centauri or Barnard's Star – literally thousands of light years – but among the 2,400 planets that Kepler did detect, several were promising candidates for habitability, based on their size and the distance from their star. Among these are Kepler 442b, which is 1,100 light years distant and orbits its star every 112 days, and since its star is a little smaller and cooler than our Sun (but bigger than a red dwarf) the habitable zone is closer in, and Kepler 442b might just be a little too cold. A better possibility is Kepler 452b, which is actually in the habitable zone of a star very similar to our Sun, orbiting once every 385 days compared to Earth's 365 days. It is 1,400 light years away, but it is much bigger than Earth, with a radius 1.6 times greater and a mass possibly five times higher. Because we don't have planets with similar dimensions in our solar system, astronomers don't really know what kind of conditions they will have.

ABOVE A rogue planet; worlds that float freely throughout the universe. These cool exoplanets don't orbit stars, going it alone on their journey. *(ESO/L. Calçada/P. Delorme/R. Saito/ VVV Consortium)*

Barnard's Star b, which is 5.9 light years away. Both are rocky worlds; Proxima b is estimated to be between 0.8 and 1.5 times the radius of Earth, while Barnard's Star b is about 1.3 times the radius of Earth. Because of the way they were discovered – by watching the dip in their star's light as the planet passes in front of the star in a so-called 'transit' – we have a good idea of their size, but not their mass. However, given their size they are unlikely not to be rocky – gaseous planets tend to be much larger.

Proxima b is in the habitable zone of its star – the region around its star where temperatures are just right, like Goldilocks's bowl of porridge, for liquid water to exist on the surface of a planet with an atmosphere. That's really exciting, because it means the closest planet to the solar system could potentially support life. There's bad news though, in the form of its star, which like Barnard's Star is a red dwarf. These, remember, are the coolest, most common and most long-lived types of stars in the Milky Way. However, for such small stars, they pack a mighty magnetic punch that leads them to release flares of charged particles as magnetic fields near their surface snap and reconnect. These flares could potentially irradiate the worlds around them, making them inhospitable. We won't really know until we have the technology to directly image these planets or visit them with spacecraft, and study them much more closely.

Searching for planets isn't all about looking for habitable worlds. There's all kinds of weird and wonderful worlds out there in the Milky Way! For example, one of the largest planets known is HD 100546b, which is 320 light years away and has a radius no greater than 6.9 times the radius of Jupiter. That's enormous – at maximum it would be over 480,000km (300,000 miles) across! There's a twist, though, that HD 100546b is still very young, just 10 million years old, and is still encased within its protoplanetary disc. It's also very hot still, and will likely contract as it ages and cools, meaning it won't stay at its current size but it will still be enormous.

The hottest known planet is Kepler-70b, which is a rocky planet smaller than Earth (a radius of 0.76 Earth radii and a mass of 0.44 Earth masses) that orbits around a star 4,200 light years away at a distance of just 900,000 kilometres (559,000 miles). It's so close to its star that its surface temperature is up to 7,000°C (12,632°F) hotter than the surface of the Sun. Its surface wouldn't just be molten, but the rock would vaporise, perhaps giving the planet an 'atmosphere' of silicates.

Another utterly bizarre world that indicates the diversity of planets that the Milky Way has to offer is 55 Cancri e. It's a large planet, with a mass 8.6 times greater than Earth and a diameter twice as great, and is part of a multi-planet system with four other worlds around a Sun-like star. However, when the Hubble Space Telescope observed the planet's atmosphere – this is achieved via a technique known as transmission spectroscopy, whereby watching the planet's atmosphere absorb some of its star's light during a transit results in identifiable absorption lines in the star's spectrum – scientists found no evidence of water vapour. This was considered highly unusual because

water is a very common molecule in the universe. Hydrogen was detected, so that must mean the planet has little to no oxygen. Instead, carbon may take the place of oxygen. On Earth, oxygen is integrated into the silicate rocks beneath our feet, but on 55 Cancri e, the rocks would be graphite instead, and the carbon would be so dense near the centre of the planet that it could have a core made entirely of diamond.

Other worlds in the Milky Way – perhaps a significant proportion, in fact – are the complete opposite of 55 Cancri e, and are rich in water. So rich, in fact, that they could have global oceans tens of kilometres deep, or thick atmospheres of water vapour. One such planet could be GJ 1214b, which is only 42 light years away. It's a super-Earth, and although water molecules have yet to be detected on the planet, its overall density strongly implies that water makes up a large proportion of its bulk. Think of a giant snowball, perhaps a giant version of one of Jupiter's icy moons, where there is a modest rocky core surrounded by layers of high-pressure water (which forms exotic types of 'ice' such as ice VII, which forms at pressures greater than 30,000 times the Earth's surface pressure, or 3 giga-pascals) with liquid water surrounding the high-pressure layers.

ABOVE An artist's impression of a super-Earth, planets that have masses higher than that of Earth but below Uranus.
(ESO/M. Kornmesser/ Nick Risinger)

ABOVE Hot, rocky planets covered in lava or with boiling oceans are just one kind of extrasolar planets discovered in the universe. *(ESO/L. Calçada)*

We mentioned earlier that between a quarter and a third of all star systems are double, triple or even quadruple (and so, if you add up all the stars in multiple systems, it turns out that most stars in the Milky Way are in multiple systems, although most systems are not multiple). Can these packed systems also have planets? It turns out that they can. One example is the three worlds orbiting the twin stars of the binary star system Kepler-47. The two stars are very close, orbiting around a common centre of mass between them once every seven days (7.45 days to be exact) with a separation of 12.5 million km (7 million miles). The three planets, which are all gas giants, then orbit both stars from further out (at distances of 43 million, 104 million and 147 million km (27 million, 65 million and 91 million miles) respectively). The outermost planet, Kepler-47d, is in the system's habitable zone, meaning that if it has any large moons, they would potentially be habitable, and get to see double sunsets (or sunrises) just like Luke Skywalker did on the planet Tatooine in *Star Wars*.

The Pleiades

One location where astronomers are finding surprisingly few exoplanets is in open star clusters. With so many stars in one location, you might think there would be planets aplenty, but so far it hasn't turned out like that. Three worlds have been discovered in the Messier 67 open star cluster, 2,500 light years away. Another trio of planets have been identified orbiting a nondescript star in the Hyades star cluster, which is located about 150 light years away. There's a smattering of other cluster planets, including a possible planet-forming protoplanetary disc around a young star, catalogued as HD 23514, in the Pleiades star cluster.

The Pleiades, more popularly known as the Seven Sisters (referring to the seven daughters of the Ancient Greek god, Atlas), is possibly the most familiar star cluster to the general public. Its seven brightest stars are named after Atlas's daughters, which are easily visible in the winter night sky in the constellation of Taurus, the Bull. Those stars are among the most massive, hottest – and therefore bluest – stars in the cluster, but although their luminosity dominates the cluster, most of the Pleiades' 1,400 or so

LEFT A familiar sight in the night sky: the Pleiades, an open star cluster in the constellation of Taurus (the Bull). Also known as the Seven Sisters, the object is 400 light years away.
(NASA, ESA, AURA/ Caltech, Palomar Observatory)

member stars are smaller, cooler stars – as we would expect based on the initial mass function.

The Pleiades will have formed in an intense star-forming region, similar to the Orion Nebula, about 115 million years ago. Astro-images show the Pleiades set against a misty blue nebula, but this is not the nebula that formed the star cluster – that nebula is long gone. Instead, it just so happens that the Pleiades are passing through an area of wispy gas.

However, there was a great deal of controversy surrounding how far away they are. Their distance is important, because astronomers use the Pleiades to calibrate their measurements of other star clusters. Because the cluster is so close, so young and so well studied, astronomers understand the properties of its stars and where they are on the HR diagram very well, but this is all highly dependent on knowing their distance. If the Pleiades were closer or further away than we thought, that would mean that their intrinsic

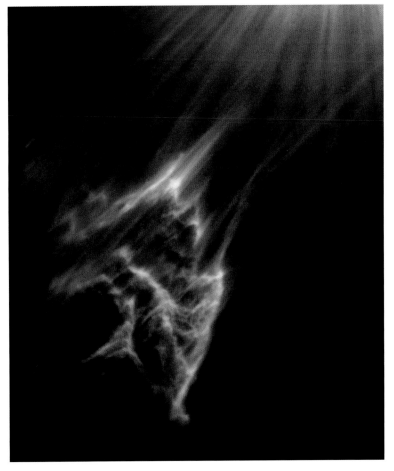

RIGHT The reflection nebula IC 349 snapped by the Hubble Space Telescope. The nebula can be found in the Pleiades. *(NASA)*

luminosities are different to what we thought in order to appear at the brightness that they do in the night sky. Different luminosities would mean we'd need to use different stellar models, and then these models would then be applied to other, more distant clusters that are too far away for their characteristics to be measured quite so well. So, getting the distance to the Pleiades wrong would cause a chain reaction of inaccuracy across the Milky Way, and even in bright clusters that we can see in other galaxies.

In 1989 the European Space Agency (ESA) launched the Hipparcos satellite, which was designed to accurately measure the positions, luminosities and distances of 100,000 stars. Astronomers were very happy with most of Hipparcos's results, but there was one bone of contention: yes, you've guessed it, the Pleiades.

Prior to Hipparcos, best estimates of the distance to the Pleiades were about 440 light years. However, Hipparcos measured them to be 390 light years away – a big difference. For nearly 25 years the Hipparcos result was strongly debated – most astronomers believed that the value of 390 light years must be wrong because the alternative was that our models of stellar evolution were skewed. Ultimately, it turned out that Hipparcos was wrong – first, in 2014 the High Sensitivity Array, which is a network of radio telescopes spread across the United States as well as including the Effelsberg 100m (328ft) radio telescope in Germany, was able to measure the parallax to the cluster. The parallax angle is how the position of an object, in this case the Pleiades, shifts against the background when we change our position, such as when Earth is on opposite sides of its orbit around the Sun, creating a baseline of nearly 300 million km (186 million miles). By knowing the baseline and knowing the parallax angle, simple trigonometry gives us a distance, which the High Sensitivity Array measured as 443 light years. Then, ESA's follow-up to the Hipparcos mission, called Gaia, repeated this measurement and also got 443 light years. The debate was over, astronomers' models of stellar evolution had been saved, and the Pleiades could continue being the gold standard for star clusters across the Milky Way, which number more than 1,000 that have been discovered, with possibly 10,000 in total in the galaxy.

Among those star clusters is the aforementioned Messier 67, which is 2,500 light years away. Whereas the Pleiades are very

RIGHT **The Double Cluster (Caldwell 14), comprising of open clusters NGC 869 and NGC 884, sits in the constellation Perseus. The pair lie at a distance of approximately 7,500 light years.** *(Wikimedia Commons)*

young, M67 is very old, with an age estimated to be between 3 and 5 billion years old. This is highly unusual; normally it is expected that the stars in open clusters will gradually drift apart and go their separate ways into the Milky Way, otherwise the Milky Way wouldn't have stars all mingling together, but just a bunch of clusters orbiting the black hole. The Pleiades are expected to disperse in about another 250 million years. Most other clusters have separated by about a billion years at the latest. Yet M67 is still hanging together, and the question is, why?

Astronomers think it is to do with the density of stars that were born within the cluster. Today it contains more than 500 stars, many of which are now expanding to become red giants, and some that have already gone through the red giant stage and are now white dwarfs. However, with it being so old, many of its most massive stars will have exploded as supernovae a long time ago. The cluster has undergone mass segregation, wherein close encounters between stars in the cluster act like gravitational slingshots, accelerating the less massive stars so that they move to the edge of the cluster or even escape it completely, while the more massive stars lose energy during the encounter and sink towards the cluster's core, emphasising the high density of stars there, the mutual gravity of which keeps the cluster together.

At the other end of the scale is the magnificent Double Cluster, which is composed of two individual open clusters in close proximity that are even younger than the Pleiades at an age of just 12.8 million years old. The two clusters, named NGC 869 and NGC 884, formed at the same time, and are located 7,500 light years away. Because they are so young, they still contain many massive stars – more than 300 massive blue supergiant stars in each cluster – that are ticking supernova time bombs.

Omega Centauri

Whereas open clusters are distributed across the galactic thin disc, another type of cluster in the Milky Way lives in the halo, which is the spherical expanse of stars and unseen dark matter above and below the plane of the spiral disc. The halo is the realm of the

ABOVE Globular cluster 47 Tucanae taken by the NASA/ESA Hubble Space Telescope. *(NASA, ESA and the Hubble Heritage (STScI/AURA)-ESA/Hubble Collaboration Acknowledgement: J. Mack (STScI) and G. Piotto (University of Padova, Italy))*

BELOW Globular cluster Omega Centauri contains up to ten million stars. *(ESO)*

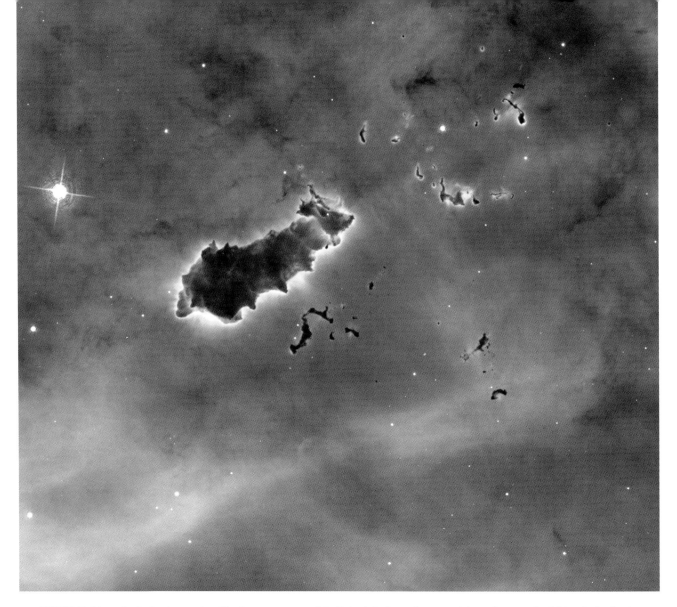

ABOVE Not too dissimilar to a caterpillar in appearance, this Bok globule glows thanks to heat and light from the hottest stars in the cluster. *(NASA, ESA, N. Smith (University of California, Berkeley), and the Hubble Heritage Team (STScI/AURA); CTIO: N. Smith (University of California, Berkeley) and NOAO/AURA/NSF)*

globular clusters, and the most mighty by far in the Milky Way is Omega Centauri.

Globular clusters get their name from their appearance: they are giant globules, a hundred light years across, filled with stars. They are invariably very old, over 12 billion years, which explains why they are not in the disc of the galaxy because they likely pre-date the disc. It's thought that the stars in each cluster formed all at once, in an incredible burst of star formation. Given that they are so old, only the lower mass stars survive, but back in their heyday when they were still filled with hot, blue supergiant stars,

they must have been astonishing to look at.

Yet Omega Centauri, which is about 15,800 light years from us, is not a normal globular cluster. For one of those we might look towards Messier 13 in Hercules, for example. Omega Centauri, on the other hand, has characteristics that suggest that it may be something else entirely: the core of a long-lost dwarf galaxy that was consumed and stripped apart by the gravity of the Milky Way.

One of the things that marks it out as special is its size: at 150 light years across it is larger than any other globular cluster in the Milky Way, and it's the second largest known globular cluster in the Local Group (the only one bigger is G1 in the Andromeda Galaxy). Within that volume, 150 light years across, are 10 million stars, all amounting together to four million solar masses, which is about ten times more massive than most globular clusters. Many of those stars are yellow like the Sun, or cooler and redder. Some are starting to make the transition to becoming red giants, and there are many white dwarfs and, presumably, neutron stars. The stars are so tightly packed that at the core of the cluster only 0.3 light years separates them on average. If the Sun and Earth were located at the centre of Omega Centauri, then the sky would be constantly ablaze with the light of all those nearby stars, making the daytime a hundred times brighter than the daytime we experience with light from just the Sun. It would be an amazing, and quite blinding, view.

Such conditions are not atypical of globular clusters – even the smaller ones pack hundreds of thousands of stars in volumes 100 light years across. What really marks Omega Centauri out as different is the ages and metallicities of its stars. If its stars had all formed at once, then they should share the same age and the same ratios of heavy elements. This is what we find in most globular clusters. In Omega Centauri, however, spectroscopic studies reveal a range of metallicities; for example, the Fe/H (iron/hydrogen) ratio is not as great in some of the cluster's stars as one might expect in a globular cluster. Furthermore, broadly speaking the cluster's main sequence stars can be split into two types – blue metal-rich stars and red metal-poor stars, with sub-groups among those two populations depending on their precise Fe/H ratio. Moreover, detailed HR diagrams of the cluster show numerous populations, where stars of the same mass and temperature seem to not be at the same stage of evolution.

So, what we seem to have with Omega Centauri is a globular cluster that has had numerous bursts of star formation, which have recycled heavy elements into the younger generations, enriching them. Unless we have misunderstood globular clusters, this is not in their nature. However, it would certainly be understandable were Omega Centauri the relic of a dwarf galaxy ripped apart. If that is true, most of its stars will now be integrated into the Milky Way. Indeed, Kapteyn's Star, which is a red dwarf 12.76 light years away, has elemental abundance ratios very similar to some of the stars in Omega Centauri, and it has been speculated that it came from there, once upon a time.

ABOVE The great globular cluster Messier 13 in the constellation of Hercules (the Hero) rests just 25,000 light years away and is 145 light years in diameter. *(ESA/Hubble and NASA)*

Further fuel to the fire of the theory that it was once a dwarf galaxy would be added if a black hole could be identified at the centre of the cluster. As we've seen, not all galaxies, particularly dwarf galaxies, contain central black holes, but some do. While black holes of modest stature have been found in globular clusters before, such as the cluster NGC 3201 which is 16,300 light years away, these are stellar mass black holes, born from supernova explosions, or perhaps neutron star mergers. However, in 2008, astronomers led by Eva Noyola of the Max Planck Institute for Extraterrestrial Physics in Germany, used the Hubble Space Telescope to measure the orbital

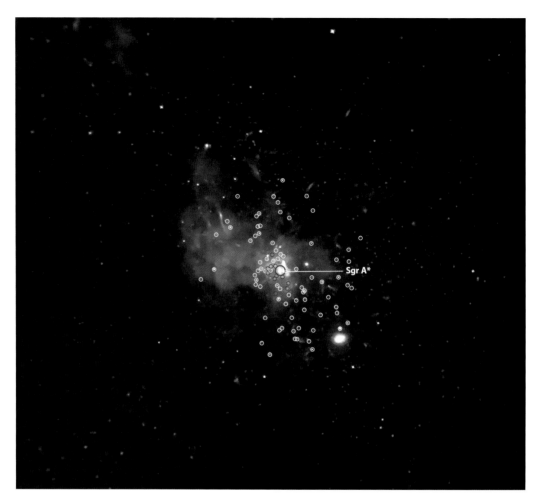

RIGHT A swarm of thousands of black holes close to the centre of the Milky Way. *(NASA/CXC/Columbia Univ./C. Hailey et al.)*

Sgr A*

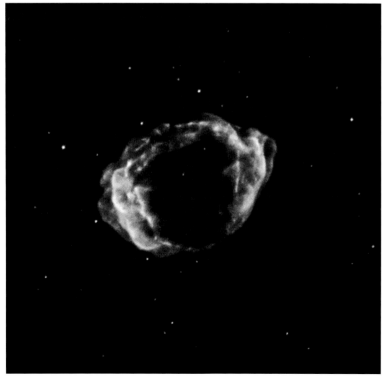

velocities of stars within Omega Centauri, and found them to be much faster than expected, based on the overall visible mass of the cluster. In particular, they found that it was the stars near the centre that were orbiting at higher velocities, as though there were some great but unseen gravitating mass there, in accordance with Kepler's laws of planetary motion. The obvious identity of such a great unseen mass was a black hole, and Noyola's team calculated its mass to be 40,000 solar masses. It would be far too big to be a stellar mass black hole produced by a supernova, yet too small to be considered supermassive like Sagittarius A*. Instead, it would be one of the first intermediate-mass black holes to be found. However, follow-up studies have

LEFT Supernova remnant G1.9+0.3, the most recent Type Ia stellar explosion in the Milky Way. *(NASA/CXC/CfA/S. Chakraborti et al)*

failed to replicate these findings, concluding that if there is a black hole there at all, it has at most a mass of 12,000 solar masses, which admittedly is still quite large. A definitive answer still awaits us.

Another mystery that is inherent not only to Omega Centauri, but many globular clusters, are the blue stragglers. These are stars that are much bluer than the other stars in a globular cluster, meaning they are hotter and therefore more massive. But if they are more massive, they should have used up their stock of hydrogen fuel long ago, yet they are still around, i.e. they appear to be 'straggling' in their evolution. What could possibly cause their youthful appearance?

Given that globular clusters are so densely packed – remember that at the centre of Omega Centauri, stars are on average just 0.3 light years apart – collisions between stars are more likely than elsewhere in the Milky Way. In particular, binary systems could easily be perturbed by the gravity of stars moving nearby, causing the two stars in the binary system to collide and merge. When that happens, they form a more massive star that is rejuvenated, mixing what hydrogen the two doomed stars had left into the core of the new star. Evidence that mergers are the cause has come from observations of blue stragglers in another large globular, called 47 Tucanae, which is about 13,000 light years away, is 120 light years in size and which may also contain an intermediate-mass black hole. Like Omega Centauri it contains multiple populations of stars and may also be a core remnant from a cannibalised dwarf galaxy. Observations with the Hubble Space Telescope of one blue straggler in 47 Tucanae in particular found that it has twice the mass of all the other stars around it – which are all evolving in unison – and is spinning 75 times faster than the Sun rotates at the equator (one rotation of the Sun at the equator takes about 24 days and 12 hours, while at the pole it is slower, taking almost 38 days). The rotation periods of stars slow down as they age, which is another indicator that a blue straggler is a new arrival on the scene, and models of stellar mergers predict that the new star that forms will be spun up.

The Zone of Avoidance

Despite the stellar collisions, and the searingly bright skies that any planet would have, one advantage of living in a globular cluster in the halo at a high galactic latitude is that they have a far better view of the Milky Way galaxy than we have. Stuck as we are in the plane of the Milky Way's spiral disc, we have to contend with looking through a cosmic ocean of dust and gas that obscures our view, dimming the light from stars and nebulae beyond, and concealing those too far away as to be too faint to penetrate all that interstellar dust. High above the spiral disc, observers in globular clusters don't have to contend with all that. They'll be able to see the overall plan of the galaxy, from how many spiral arms it really has to how large its bar is, and what is on the other side, beyond the Zone of Avoidance.

The Zone of Avoidance is the region on the night sky obscured by the densest parts of the Milky Way. Objects on the other side of this Zone are very faint, or can't be seen at all. Infrared and radio observations can see through some of it, but the simple fact is that we can really only see half of our galaxy. The other half, behind the Zone of Avoidance, is something of a mystery. There may be supernovae exploding, for example, on the far side of the galaxy that we can never see. It also obscures about 20 per cent of the rest of the universe – there are entire galaxies and massive galaxy clusters hidden behind the Zone of Avoidance; for example, something with a huge gravitational pull, known as the Great Attractor, lies beyond the Zone of Avoidance, its gravity pulling the Milky Way and the rest of the Local Group, and even the Virgo Cluster, inexorably towards it. We can't see for sure what this massive object is, but we know it must be some kind of giant galaxy cluster.

So half our galaxy will forever remain hidden from us, unless we can develop some form of interstellar space travel in the future that will take us to a better vantage point. Who knows what fabulous nebulae and star clusters reside on the far side of the galaxy? It seems a shame that there is so much about the Milky Way that we will never know.

Chapter Seven

The future of the Milky Way

Many would settle for growing old gracefully, but that's not the Milky Way's style. Its most dramatic days are still in front of it. With galaxy collisions and even a change of shape on the menu, what is going to happen to our home galaxy?

OPPOSITE Locked in a galactic embrace, the two galaxies NGC 4038 and NGC 4039 form the Antennae Galaxies – the result of a collision between two spirals like our very own galaxy. The violent crash has caused a violent outburst of young stars, shown here in flashes of blue. *(ESA/Hubble & NASA)*

It's been known since 1912 that the Andromeda Galaxy – then thought of as the 'Andromeda spiral nebula' – is moving towards us. It was around this time that the astronomer Vesto Slipher of Lowell Observatory in Arizona, made the first red shift measurements of what we know today to be galaxies exterior to the Milky Way. Red shift describes the stretching of the wavelength of light towards redder wavelengths as the object emitting that light moves away from us, whereas if an object were moving towards us, the wavelength of its light would be shortened towards bluer wavelengths. Slipher found, somewhat puzzlingly, that almost all of the spiral nebulae exhibited red shift, with one notable exception being the Andromeda spiral, the light of which was blue shifted.

The consequences of this weren't fully recognised at the time. After all, in 1912 it was still thought that the spiral nebulae were contained within the Milky Way. Only after Edwin Hubble showed us the truth did reality hit home: the Andromeda Galaxy is moving towards the Milky Way, and one day they are going to collide.

There's no need to panic just yet. We've got a long wait on our hands – at least 4.5 billion years and possibly longer. In fact, the Sun may have expanded into a red giant, swallowing Mercury, Venus and Earth in the process, and transformed into a white dwarf before the merger is all over and done with – the Sun has a lifespan of about 10 billion years, and it's already 4.6 billion years into that.

Mergers in the universe

Major galaxy collisions are not a new phenomenon. They've been happening to galaxies throughout cosmic history. In fact, they used to be far more common in the first few billion years after the Big Bang, when cosmic expansion had not yet taken galaxies far away from each other. That said, there are still plenty of examples in the local universe, and by studying them, we can piece together what fate might have in store for the Milky Way and the Andromeda Galaxy.

When we look at galaxies, and galaxy mergers, in the universe, we are seeing them as they exist at one moment in time. The timescales of looming collisions are so huge that mergers take hundreds of millions of years to proceed, so we only ever see snapshots. However, by studying these snapshots of different galaxies at the different stages of their mergers, we can get an idea of the overall picture.

First, there is the close approach. This is when galaxies initially move into each other's gravitational spheres of influence. We can see examples such as the Arp 256 system, which is a pair of interacting galaxies about 350 million light years away. The spiral structure of each galaxy is becoming distorted due to the gravitational tidal effects emanating from each one. Tidal effects occur when those parts of the two galaxies closest to one another begin to tug gravitationally, pulling themselves close

RIGHT An artist's impression of what is likely to be observed from Earth when the Milky Way and Andromeda collide. *(NASA; ESA; Z. Levay and R. van der Marel, STScI; T. Hallas; and A. Mellinger)*

to one another. Meanwhile, the parts of each galaxy on opposite sides don't feel the same amount of gravitational force, leading to a gravitational gradient across the two galaxies that begins to distort them, stretch them out and start to pull gas and stars out of them.

Indeed, the spiral arms of the galaxies in the Arp 256 system appear to be unravelling, producing streams of stars that are either trailing behind them, or reaching out to the other galaxy. But that's not the only sign that a merger is on the cusp of occurring. The two galaxies are described as being a 'luminous infrared system' – they're undergoing a huge amount of star formation as the gravitational tides stir up the giant molecular gas clouds, causing them to form stars at a frenzied rate.

The next step seems to vary. Sometimes the galaxies will crash directly into one another, but often they're a little off target and make a first close passage instead, perhaps just 100,000 light years or so apart. This is when we get examples like 'The Mice', which is the nickname for the interacting galaxy pair otherwise known as NGC 4676. They're called The Mice because of their long tails of gas and stars that have been pulled out of the two galaxies by gravity and extend for hundreds of thousands of light years. Such tails are called 'tidal tails' and are the result of the same gravitational tides as described above, as gravity pulls differently on different regions of each galaxy. The tidal tails are made of stars and gas pulled out of each galaxy, but in some instances, stars have even been observed forming within tidal tails.

The Mice are 300 million light years away, and computer simulations suggest that we are seeing the two galaxies about 160 million years after their first close passage. It won't be their last.

The Mice will swing back around, forever now in each other's gravitational grasp. It's a galactic waltz of epic proportions. They'll then collide and merge, or make another close passage before swinging back around once again, as though they were attached to elastic bands, to finally merge.

We can see what this looks like in the well-studied example of the Antennae Galaxies (NGC 4038/39). They look like one tortured, misshapen galaxy, but there are actually two galaxies. Like The Mice, they have enormously long, curving

'RED AND DEAD' GALAXIES

Elliptical galaxies are referred to as 'red and dead' galaxies. Following their formation via the merger process, the resulting bursts of star formation are so great that they rapidly use up the remaining molecular gas. Meanwhile, during the merger process the supermassive black holes that sat at the centre of each galaxy slowly sink to the core of the new galaxy, swallowing huge amounts of matter that has been disturbed by the gravitational tides. This accretion onto the black holes – which eventually merge into one – produces feedback from the hot accretion disc, which heats the molecular gas, preventing it from forming stars and ejecting it from the galaxy. The star formation and the feedback use up all the gas quickly, and then begins the process of stars formed in the starburst dying as supernovae, blowing vast amounts of dust into interstellar space. Soon, with no new stars forming, only the longer-lived red stars are left, hidden beneath blankets of dust in the form of stellar 'ash' from supernovae.

During the merger, the ordered orbits of stars in the two spirals become disordered, with stars flung onto random orbits at all angles. This leads to the bloated, elliptical shape of the newly formed 'phoenix' galaxy. We can see instances of elliptical galaxies in the shape of Messier 86 and Messier 87 in the Virgo Cluster, for example.

tidal tails that look like insect antennae, hence the system's name, but they're a lot further along in the merger process, entwined and entangled with one another, their galactic magnetic fields all twisted up, their clashing clouds of gas and dust sparking bursts of star formation that appear like blisters on their skin, and the subsequent rash of supernovae that burst out of those blisters. Five supernovae have been reported in the Antennae Galaxies over the past 100 years, the most recent being in 2013.

Where the two galaxies begin and end has become lost in the maelstrom. Once upon a time, NGC 4038 was a barred spiral similar to the Milky Way, while NGC 4039 – originally the larger of the two – was a plain spiral galaxy. Then, computer simulations suggest that 900 million years ago they began their deadly dance and began moving towards an embrace with another, before performing their first close pass 600 million years ago, causing tidal tails of stars and gas to stream out. Then they turned back and fell back onto one another, initiating the process of coalescence. The simulations imply that in 400 million years' time, the merger will be complete, with the result being a giant elliptical galaxy.

Making the Milky Way normal again

Before the Milky Way's date with destiny, it has another interloping galaxy to contend with. The Large Magellanic Cloud (LMC) is a relatively recent arrival in the Milky Way's neighbourhood and had been thought to just be passing by. However, measurements of its total mass have discovered that the LMC contains more dark matter than had been previously thought, and that changes everything.

With that greater mass comes a greater level of dynamical friction with the Milky Way. Astronomers at the University of Durham have calculated that in about a billion years' time, this will cause the LMC to lose so much orbital energy that it will slow and reverse course, heading straight for the centre of the Milky Way, with which it will merge in about 2.4 billion years' time.

According to the Durham astronomers, the collision between the Milky Way and the LMC will straighten a lot out that's wrong with the Milky Way. For one thing, for a galaxy its size, Sagittarius A* is too small according to a correlation that has been observed in other galaxies regarding black hole mass and the mass of bulge or inner region of the galaxy. Another problem with the Milky Way is that

although the halo is meant to be less chemically evolved than the spiral disc, it is even less abundant in heavy elements than the haloes of other similarly sized spirals like Andromeda. It's also less massive than them too. Finally, only 10 per cent of other spiral galaxies have a close neighbour as bright and massive as the LMC. Why is the Milky Way so special?

The Durham astronomers just think that the Milky Way is a slow maturer, and that the LMC is the growth pill that it needs. For ten billion years the Milky Way's disc has been able to evolve secularly, without interruption from collisions with other galaxies, but such a peaceful existence seems to go against the norm for most spiral galaxies that do endure such mergers from time to time. The implication therefore is that other galaxies may have had their own LMC-type companions once upon a time, but merged with them long ago, giving them the characteristics that they exhibit today and which our Milky Way apparently lacks. The Durham astronomers therefore see our impending merger with the LMC as being long overdue.

Their simulations indicate that the collision and subsequent merger will kick-start activity in the central Milky Way. Sagittarius A* will dramatically switch on and gas from both our galaxy and the LMC will find its way ushered down the throat of the supermassive black hole, and estimates suggest that Sagittarius A* would be fattened up by all that material, growing in mass eight times over and potentially releasing jets of particles moving at close to the speed of light, accompanied by powerful gamma-ray radiation. Meanwhile, as the LMC is disrupted and ultimately torn apart by the gravity of the Milky Way, most of its stars will be subsumed into the Milky Way's halo. Because the metallicity of the LMC's stars is on average higher than the metallicity of the Milky Way's halo, the LMC stars will chemically enrich the halo, and increase the halo's mass by between three and six times over. Some stars from the Milky Way's disc could also be swept up in the disorder caused by the merger and ejected out into the halo, but this is expected to be but a small fraction of the total number of stars in our galaxy. Although there's a chance that the Sun and its solar system could be kicked out into the halo, the odds are in our favour that it won't happen.

BELOW Our Sun would have used up all of its fuel and swollen into a red giant before the titanic collision occurs. Most of the terrestrial planets – including Earth – would have been destroyed. *(ESO/S. Steinhöfel)*

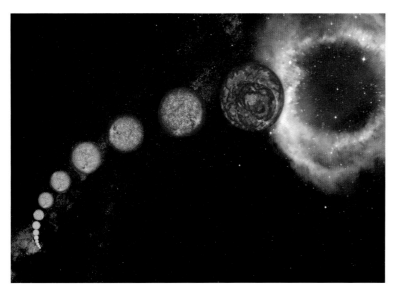

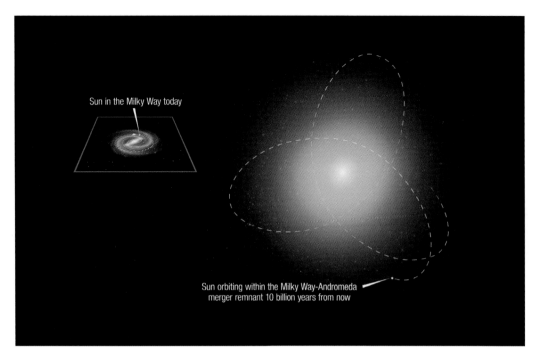

Sun in the Milky Way today

Sun orbiting within the Milky Way-Andromeda
merger remnant 10 billion years from now

The Milky Way–
Andromeda Collision

While the merger with and subsequent cannibalisation of the LMC will be significant, it's just a teaser for the main event, which is the forthcoming collision with the Andromeda Galaxy. However, the LMC merger is going to have an effect, because it is going to make the Milky Way more massive; as we have seen in previous chapters, Gaia and the Hubble Space Telescope measured the mass of the Milky Way as 1.5 trillion solar masses, whereas the mass of the LMC has been measured as 138 billion solar masses. This extra mass will affect the Milky Way's motion through space, which the Durham astronomers suggest will, in turn, see the collision with the Andromeda Galaxy happen later than expected, 5.3 billion years from now.

However, there is some uncertainty in this estimate, because until recently it has only been possible to measure the line-of-sight velocity with which the Milky Way and the Andromeda Galaxy are approaching one another. This is done the old-fashioned way, the way that Vesto Slipher did it in 1912, by measuring its blue shift. This is done spectroscopically, splitting the light into its component wavelengths using a diffraction grating, and then identifying the

positions of well-known absorption or emission spectral lines before comparing them to their spectral lines when at rest in the laboratory, to see how much they have Doppler shifted by. The degree of blue or red shift correlates with a specific line-of-sight velocity. For the Andromeda Galaxy, this has been measured to be an approach velocity of 110km (68.3 miles) per second. It might not sound like much when compared to the enormous scales of each galaxy and the intergalactic space between them, but it is fast enough that over billions of years the two galaxies can close down the 2.5 million light years (23.7 million trillion km or 1.47 million trillion miles) between them.

What was not so easy to measure was Andromeda's lateral motion relative to the Milky Way (all of these measurements are relative to the Milky Way, making the assumption that we are at rest to make things simpler, but in truth part of that 110km per second is provided by our Milky Way moving towards the Andromeda Galaxy too). It can't be measured by Doppler shift because it doesn't cause any stretching or compression of the wavelengths of light, and the Andromeda Galaxy's motion on the sky (what we call its 'proper motion') is very small and extremely difficult to measure. What is needed is a highly sensitive astrometric experiment that can measure the minutest changes in position on the sky. Step forward Gaia, once again.

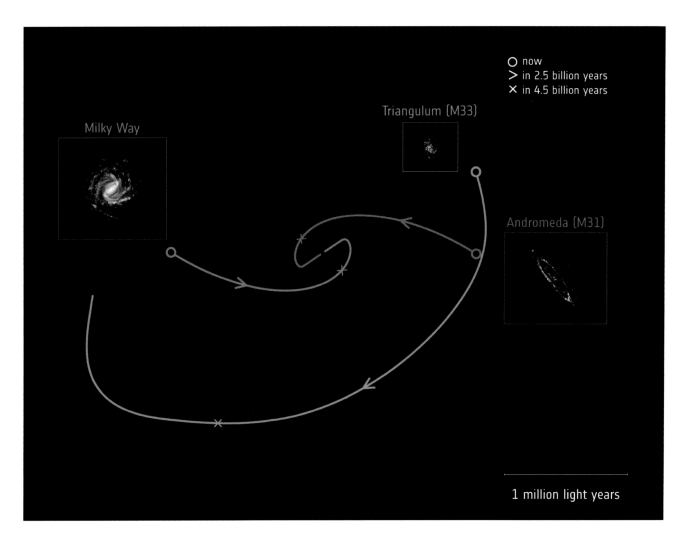

Milky Way

Triangulum (M33)

○ now
＞ in 2.5 billion years
✕ in 4.5 billion years

Andromeda (M31)

1 million light years

ABOVE The future orbital paths of Andromeda, the Milky Way and the Triangulum Galaxy – the latter of which will collide with Milkomeda after orbiting the galactic smash-up. *(Orbits: E. Patel, G. Besla (University of Arizona), R. van der Marel (STScI); Images: ESA (Milky Way); ESA/ Gaia/DPAC (M31, M33))*

Gaia is sensitive enough that it is able to measure the individual motions of thousands of bright stars, not just in the Andromeda Galaxy, but in the LMC and the Triangulum Galaxy too. Analysis of this data collected by Gaia indicates that the Andromeda Galaxy's lateral velocity is 57km (35.4 miles) per second. In other words, while it is moving towards us at 110km per second, it's also moving off to the side at 57km per second. Could this be enough that the Andromeda Galaxy will miss us?

Not quite, but it does mean that the collision will probably not be a direct head-on smash. Instead, the Andromeda Galaxy will first make a close pass, perhaps a glancing blow. When exactly this will occur is not clear; estimates of the timescales involved have always varied. The Gaia results suggested this might happen in 4.5 billion years' time, but that does not take into account the collision with the LMC, which the Durham team suggest

will delay the Andromeda–Milky Way smash until 5.3 billion years from now, but that in turn doesn't take into account the lateral motion of the Andromeda Galaxy. On top of that, the Triangulum Galaxy may also have a role to play – it's heading towards the Andromeda Galaxy, and its gravity could slow the Andromeda Galaxy down.

If it is, as looks increasingly likely, a close first passage between the Milky Way and the Andromeda Galaxy, then whenever it happens, we can expect a scenario similar to The Mice. Our two galaxies will start ripping chunks out of each other, tidal streams trailing away, their central bars destroyed and their spiral arms twisted this way and that. Perhaps there will be a second close encounter after that as they spiral towards each other, and then, in this deathly gravitational embrace, the two disrupted galaxies will ultimately spin back around and literally fall into each other's arms,

162
MILKY WAY MANUAL

WHAT WILL HAPPEN TO THE SUN?

We've already seen that there's a slim chance that the LMC collision will knock the Sun into the Milky Way's halo, but the impending giant crash with the Andromeda Galaxy is definitely not something that the Sun can avoid.

Despite all the stars involved in the collision, we don't have to worry about the Sun colliding with other stars (assuming that the Sun is still around at this point in the future). The average distance between stars, at least in the solar neighbourhood – and this is thought to be the case all over the galaxy's disc – is about 4 light years or 37 trillion km (23.5 trillion miles). The diameter of the Sun is just 1.39 million km (863,705 miles) and the diameter of the solar system out to, say, the orbit of Pluto, is about 15 billion km. The scale of the Sun and the solar system is tiny compared to the huge distances between stars, so there is plenty of room for all of those intermingling stars to pass through the wide gaps between each other without colliding. Indeed, the entire galaxy merger will probably proceed without a single collision between two stars.

their stars and gas becoming mingled. Their great molecular gas clouds will meet head-on and huge bursts of star formation will ignite. Over hundreds of millions of years, our two galaxies will coalesce into a new giant elliptical galaxy, that has been flippantly nicknamed 'Milkomeda'. Regardless of the name, it will be the greatest transformative event in the history of the Milky Way.

However, the gravitational tides that will wash over both galaxies during the process of first the close passages and then the merger will scatter stars like skittles. Astronomers T. J. Cox and Avi Loeb of Harvard University have run computer simulations that point to there being a 67% chance that the Sun and the solar system will be pushed to the outer suburbs of Milkomeda at a distance greater than 65,000 light years from the centre. For comparison, our Sun is currently about 26,000 light years from the centre of the Milky Way. There's even a

ABOVE It's likely that the resulting collision will form a giant elliptical galaxy. However, astronomers haven't ruled out the possibility of a lenticular – of which the Spindle Galaxy (NGC 5866) in Draco is a prime example. The deciding factor will be the amount of gas that remains. *(NASA, ESA and the Hubble Heritage Team (STScI/AURA))*

BELOW When black holes merge, they create ripples in space-time known as gravitational waves. *(Swinburne Astronomy Productions)*

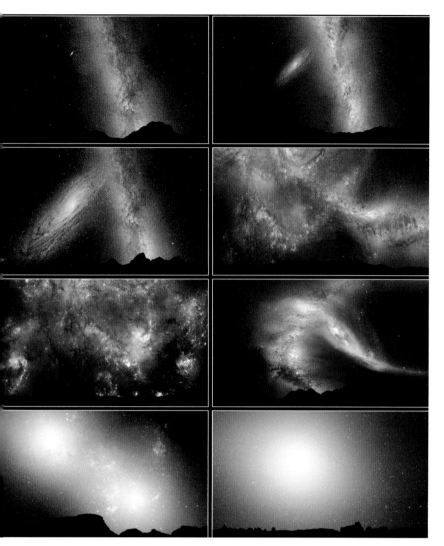

LEFT A full sequence of how the collision between the Milky Way and Andromeda is likely to play out. In two billion years, our closest spiral's disc will start to appear larger in the night sky. The end result: Milkomeda. *(NASA; ESA; Z. Levay and R. van der Marel, STScI; T. Hallas; and A. Mellinger)*

12 per cent chance that the Sun could be ejected from Milkomeda completely, left to spend eternity drifting lonely through intergalactic space. Alternatively, Cox and Loeb calculate that during the second close passage between the Milky Way and the Andromeda Galaxy there is a 3% chance that Andromeda could steal the Sun and the solar system from the Milky Way, plucking us out with its gravity. Of course, we'd only be swapping galaxies for a relatively short while, as the eventual merger will bring us back together again.

Black hole crash

Meanwhile, a cataclysmic event will be unleashed in the core of Milkomeda. The Milky Way's supermassive black hole, Sagittarius A* – which should have grown in mass quite significantly by this point to over 30 million solar masses thanks to the Milky Way's earlier merger with the LMC – will be sinking to the core of Milkomeda at the same time that the

RIGHT When two galaxies collide, their clouds of gas and dust also crash into each other – on some occasions, you get a starburst galaxy, just like NGC 1569. *(NASA, ESA, the Hubble Heritage Team (STScI/ AURA), and A. Aloisi (STScI/ESA))*

Andromeda Galaxy's supermassive black hole, which contains 60 million solar masses, is also sinking towards the heart of Milkomeda. Eventually they are going to come into each other's sphere of influence, and then the fireworks will really begin.

As they lunge through the burgeoning bulk of Milkomeda, they'll each start gobbling up a lot more interstellar gas, growing enormously. As they do, and as they gravitationally interact with stars that they encounter, dynamical friction will cause orbital energy to be transferred from the black holes to the stars. The stars will get pushed outwards to wider orbits as they gain orbital momentum, while the black holes will lose it and sink closer inwards. Over millions of years and countless encounters with stars (like the stars, the black holes are relatively small compared to the distances between the stars, so again it's more likely that the gravity of the black holes will nudge stars from a distance onto different orbits – effectively the stars will receive a gravitational slingshot – rather than the stars colliding with and being consumed by the black holes) will see the two black holes make their way to the new core of Milkomeda.

We can get an idea of how this might go down from the merging galaxy system NGC 6240, which is estimated to be 400 million light years away. NGC 6240 is a bit of a mess, with loops and streams of tidally ripped-out stars all over the place, a completely disrupted shape, and thick clouds of dust concealing hurried bouts of star formation. The two spiral galaxies that originally merged appear to be completely gone, but despite all this, we can see the double nuclei at the centre of NGC 6240, made up of two supermassive black holes, one from each galaxy. They are only 3,000 light years apart and will merge within the next 10 or 20 million years.

NGC 6240 is what astronomers term an 'ultraluminous infrared galaxy', or ULIRG for short. Such galaxies are exactly what they are described as – objects that shine surprisingly brightly at infrared wavelengths. The infrared light is emitted thermally by interstellar dust that has been heated up. The identity of the heat source remains uncertain, not just in NGC 6240, but in all ULIRGS. It's either produced by radiation from hot, massive stars that are forming in their hundreds in the starbursts, or

ABOVE Quasars, as shown in this artist's impression, possess very bright centres that are powered by supermassive black holes. It's possible that, when the two galaxies' centres collide, they could form an Active Galactic Nucleus (AGN). *(ESO/M. Kornmesser)*

LEFT An artist's impression of an Active Galactic Nucleus (AGN), powered by a supermassive black hole gorging on gas and dust at an extraordinary rate. *(ESA/NASA, the AVO project and Paolo Padovani)*

LEFT An Active Galactic Nucleus (AGN) inside the galaxy Centaurus A. The outflows are clearly visible. *(ESO/WFI (Optical); MPIfR/ESO/ APEX/A. Weiss et al. (Submillimetre); NASA/ CXC/CfA/R. Kraft et al. (X-ray))*

by outflows from active supermassive black holes, or both. The trouble is that all that dust hides much of what is going on inside merging galaxies. There is even a suggestion that merging galaxies could produce quasars – these are the most active, most brilliantly luminous active galactic nuclei – but that the dust is hiding the presence of the quasars.

NGC 6240 provides us with one of our best views into what is happening inside a ULIRG. The Chandra X-ray Observatory has detected powerful X-rays coming from the double nuclei within NGC 6240, indicating that they are both highly active, chomping down on an accretion disc of extremely hot gas surrounding each black hole, and radiating away tremendous amounts of energy from that accretion disc. Whether they are bright and active enough to be described as a quasar has not yet been determined.

As the two black holes within NGC 6240 inch ever closer to each other, the winds of radiation that they expel strengthen. Astronomers, led by Francisco Müller-Sánchez of the University of Colorado, Boulder, have observed two outflows of radiation and energetic particles emanating from the system's double nucleus. One outflow, evidently from one or both of the black holes, pushes 75 solar masses worth of gas out of the centre of NGC 6240 each year, while a second outflow containing ten solar masses per year is made by the hot massive stars, produced by a burst of star formation very close to the double nuclei, showing that the chaos around the black

holes is conducive to rapid star formation. On the other hand, the mass contained in the outflows, which are able to escape NGC 6240 completely, is of a similar amount to the star formation rate in NGC 6240, which is about 80 to 90 solar masses per year. This implies that the outflow is providing negative feedback, removing potential star-forming material at the same rate, or faster, than which stars themselves are forming. It's the beginning of the end, and as all the star-forming molecular gas is used up in NGC 6240, or expelled from its confines, the galaxy, which will evolve into an elliptical, will become red and dead, just like all the others.

So, we can expect a very similar process to occur in Milkomeda, and the remnants of the Milky Way and the Andromeda Galaxy may very well switch on to what might be considered borderline quasar activity.

Then, once the black holes get within one light year of each other, they will start to emit significant gravitational waves. These ripples in space-time are simply produced by the huge masses of the two black holes as they accelerate in each other's respective gravitational fields as they orbit one another. The energy for the gravitational waves comes from the orbital energy of the black holes around each other. So gravitational waves remove orbital energy from a system, causing the black holes to spiral closer and closer, and in turn to orbit faster and faster in order to conserve their angular momentum, just as a figure skater spins faster when they pull their arms in. The gravitational waves increase in frequency as the black holes' in-spiral becomes tighter and tighter until they are so tight that they merge. Some of their mass-energy will be carried away by the gravitational waves, but it is entirely possible that by this stage Sagittarius A* and the Andromeda Galaxy's black hole will have individually grown so massive that when they merge they will easily form a new black hole with over 100 million solar masses.

With this final step, the transformation of the Milky Way and the Andromeda Galaxy into the giant elliptical galaxy that we call Milkomeda will be complete, signalled by that final 'chirp' of gravitational waves as the black holes coalesce. It will be a long, drawn-out affair, possibly taking up to, or even over, a billion years between the first close passage between

BELOW An Active Galactic Nucleus (AGN) can release as much energy as 100 million supernova explosions. *(NASA, ESA, and J. Olmsted (STScI))*

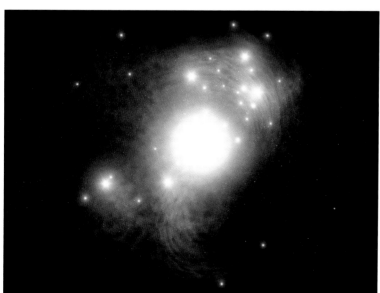

the galaxies and the black hole merger. One might imagine how, in the far, far future, some alien civilisation living in a distant galaxy, armed with their own gravitational wave detectors, might detect a faint burst of gravitational waves that have travelled across the universe – the echo of the two mighty supermassive black holes coalescing, and the final statement on the evolution of the Milky Way.

Not that it will be the end of the collisions. At some point the Small Magellanic Cloud will follow its brother and collide and merge with either the Milky Way or Milkomeda. The Triangulum Galaxy doesn't escape scot-free either; currently falling towards the Andromeda Galaxy, it will be dragged along in the Andromeda Galaxy's wake and eventually be subsumed by the newly formed Milkomeda. Briefly, the enormous black hole at the centre of Milkomeda will reawaken as material from the Triangulum Galaxy is consumed, but once that has run out, the black hole will finally fall into permanent quiescence.

With no more molecular gas, star formation will stop within Milkomeda. As the aeons go past, the remaining stars will gradually die off, expanding into red giants that then puff off their bloated outer layers to leave behind white dwarfs. The smallest stars – the red dwarfs – can last for trillions of years, but eventually these too will expire. All that will remain will be dead husks orbiting around the supermassive black hole: brown dwarfs that were too small to ever turn into stars, dead rogue planets, dark neutron

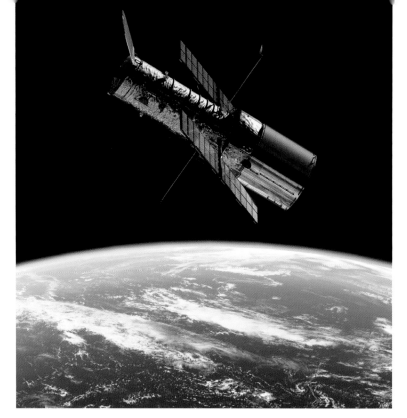

stars and white dwarfs slowly cooling down to become black dwarfs as they equalise with the background temperature of the universe, just a few degrees above absolute zero: minus 273.15°C (minus 459.67°F). Over hundreds of trillions of years, matter itself will break down, atoms will decay, while Milkomeda's remaining black hole will begin to evaporate through a process discovered by Stephen Hawking, and named Hawking radiation. Eventually, 10^{100} years from now, every part of the cosmos that we recognise today will disappear and our wonderful Milky Way galaxy will be but a faded memory.

ABOVE The Hubble Space Telescope measured the motions of stars in Andromeda, concluding that there is 100 per cent certainty that the collision will occur. *(ESA)*

LEFT These two galaxies, IC 2163 (left) and NGC 2207 (right), are in the process of a merger, creating a tsunami of star formation and likely to reflect the fate of the Milky Way and Andromeda. Supernovae have been observed in the collision. *(ALMA (ESO/NAOJ/ NRAO)/M. Kaufman)*

Glossary

Bar – A rectangular-shaped structure of stars and gas running through the centre of some spiral galaxies, thought to form as a result of gravitational instabilities in the spiral disc.

Black hole – A point of infinite mass and density, wielding enormous gravitational forces. Supermassive black holes, which are millions or billions of times more massive than the Sun, are found at the centres of galaxies.

Bulge – The central region of a classical spiral galaxy, so-named because it appears to bulge out of the spiral disc. Bulges tend to contain older stars, although they have been found to contain younger stars too.

Cepheid variable – A pulsating star that undergoes periods of brightening and dimming as it pulses. They have a period-luminosity relation, whereby the longer the period, the brighter the star. By comparing this intrinsic brightness with how bright or faint they appear in our sky, we can calculate how far away they are and use them as distance markers across the Universe.

Constellation – A pattern of stars in the sky, often named after mythological characters such as Orion the Hunter or Pegasus the Winged Horse. Although the stars appear close together in our sky, in reality they may be dozens, hundreds or thousands of light years apart.

Cosmic Microwave Background (CMB) radiation – The relic radiation from the Big Bang, which has cooled throughout cosmic history to just 2.7 degrees above absolute zero, detectable at microwave wavelengths.

Dark matter – A mysterious substance that apparently makes up most of the mass of galaxies, based on the extra gravity in galaxies and galaxy clusters that can only be explained by some hidden, 'dark' matter. What dark matter is made from is unknown.

Dwarf galaxy – A smaller galaxy compared to the likes of our Milky Way Galaxy, with a diameter of a few thousand light years, and often found in orbit around larger galaxies.

Elliptical orbit – An orbit in the shape of an ellipse rather than a circle. The mathematics behind the shape of these orbits was calculated by Johannes Kepler.

Exoplanet – A planet orbiting a star other than the Sun. Statistical evidence suggests that on average every star in the Galaxy has at least one exoplanet orbiting it.

Galaxy – A collection of up to a trillion stars, as well as copious amounts of interstellar dust, gas and dark matter, all gravitationally bound.

Galactic habitable zone – The region in a Galaxy where conditions are most conducive for the formation of potentially habitable planets.

Giant molecular cloud (GMC) – An enormous, lumbering cloud of molecular gas, mostly hydrogen, with a mass thousands of times greater than the Sun. They are the birthplaces of stars.

Globular cluster – A ball of ancient stars, around a hundred light years across, but containing hundreds of thousands, or even millions of stars. They are found in the haloes of galaxies.

Halo – A sparse region of old stars and low metallicity that surrounds a galaxy.

Hertzsprung-Russell Diagram – A graph, charting the luminosities of stars against their temperature (sometimes colour of the star is used as a proxy for temperature). Where a star is on the diagram tells us about what stage of its evolution it is at.

Hubble Tuning Fork – A system of classifying galaxies in the shape of a tuning fork, with barred spirals on one prong, ordinary spirals on the other, and ellipticals on the 'handle'.

Hydrogen – The simplest element, consisting of just an electron orbiting a proton. Formed in the Big Bang, it is the most common element in the Universe. Atomic hydrogen radiates at a wavelength of 21cm, while molecular hydrogen is able to form stars.

Kepler's laws of planetary motion – All planets move around the Sun in elliptical orbits, a line between a planet and the Sun sweeps out equal areas inside its elliptical orbit in equal time, and the square of the orbital period is proportional to the cube of the orbit's semi-major axis (half of the long axis of the ellipse).

Light year – The distance that light, moving at 299,792,458m (984 million ft) per second, in one year, which is 9.46 trillion km (5.88 trillion miles).

Local Bubble – A region of less dense gas in the interstellar medium through which the solar system is passing.

Local Group – The nearby collection of about two dozen galaxies, including the Milky Way Galaxy, the Andromeda Galaxy and the Triangulum Galaxy.

Magellanic Clouds – Two nearby dwarf galaxies that are easily seen in the sky from the Southern Hemisphere. They are individually named the Large and the Small Magellanic Cloud.

Metals – Astronomers (but not chemists) refer to elements heavier than hydrogen and helium (which were produced by the Big Bang) as 'metals'. These elements are formed inside stars and the metal content of the Universe has built up over time.

Milkomeda – The product of the impending collision between the Andromeda Galaxy and our Milky Way Galaxy.

Galaxy mergers – Galaxies often collide with one another. A minor merger involves a small galaxy colliding with a large galaxy like the Milky Way, while a major merger is a collision between two large galaxies, forming an elliptical galaxy.

OB Association – A loose collection of the hottest and brightest stars, which are frequently the source of supernovae.

Orion Arm – The small spiral arm of our Milky Way that the solar system is passing through.

Planetary nebula – When a Sun-like star reaches the end of its life, it sheds its middle and outer layers, which create an expanding nebula called a planetary nebula.

Population I and II stars – Stars are split into two 'populations' depending on their metallicity. Population I stars, such as the Sun, are younger and richer in metals, whereas older stars with fewer metals are referred to as Population II.

Pseudobulge – Some spiral galaxies that display no evidence for significant galactic mergers don't have a classical bulge, but a smaller, boxy bulge called a pseudobulge,

which contains stars of similar ages and metallicities as the disc.

Quasar – A distant galaxy with an active black hole that is beaming a jet of radiation in our general direction, making the galaxy – or rather, its active galactic nucleus – appear bright.

Redshift – The Doppler effect causes the light from objects that are moving away from us to become shifted towards redder wavelengths. The faster a galaxy, for example, is receding from us, then the greater its redshift, which is correlated to its distance.

Sagittarius A* – The name given to the 4.1 million solar mass supermassive black hole at the centre of our Milky Way Galaxy.

Spiral arms – Curving arms of gas and dust that wrap themselves around the discs of spiral galaxies. The material within the arms is not fixed, but is just passing through, implying they are more like density waves than rigid structures.

Starburst – A galaxy that is undergoing an enhanced level of star-formation, often brought about by gravitational disturbances from close encounters with other galaxies, which stirs up molecular gas to the point that it can form stars in a frenzy.

Supernova – Either the explosion of a massive star with at least eight times the mass of the Sun, or the destruction of a white dwarf star that has accumulated enough matter from a companion star (or merged with another white dwarf) to exceed the critical mass limit of 1.4 solar masses.

Thick disc – The oldest, and thickest, part of the Milky Way's spiral disc. It is about 1,000 light years deep and is filled with stars mostly older than 10 billion years.

Thin disc – The narrower, younger part of the Milky Way's spiral disc, which exists within the thick disc, and contains stars much younger than 10 billion years.

Tidal streams – When two galaxies interact gravitationally, tidal forces can pull out stars and gas from a galaxy, flinging them collectively as streams, thousands of light years long, into intergalactic space.

Winding dilemma – If spiral arms were rigid structures, then as a galaxy rotates, the arms should wrap around it. The fact that they don't is called the winding dilemma, and is solved by considering spiral arms as density waves instead.

Index

General

Alpher, Ralph 29
Ancient Greek mythology 98, 105, 148
Apache Point Observatory, New Mexico 42, 118
Apollo 17 9
Arecibo Galaxy Environment Survey (AGES) 67
Aristotle 98
Armstrong, Benjamin 71
Astrobiologists 111
Atacama Desert 118
Atmospheric seeing 110
Australian National University 70

Baade, Walter 40-41
Barnard, Edward Emerson 129
Beaton, Racheal 61
Bell, Jocelyn 138
Bell Labs, New Jersey 29
Blecha, Laura 63
Bosma, Albert 42
Brahe family 17
Brahe, Tycho 16-18, 141
    death 17
Briggs, Franklin 43, 45

Cambridge Catalogue of Radio Sources 25
Cardiff University 66-67
Carey Larkin 120
Carnegie Mellon, University, Pennsylvania 103
Center for Physical & Technological Sciences, Lithuania 50
Cerro Tololo Inter-American Observatory (CTIO), Chile 36
    Bulge Radial Velocity Assay (BRAVA) 35-37
Chen, Xiaodian 42-43, 45
Chinese Academy of Sciences, Beijing 42
Chinese astronomers 139
Clarkston, Will 38, 100
Copernicus, Nicolaus 16, 77
Cox, T.J. 163-164
Curtis, Heber 14-15

Dame, Tom 39
Dékány, Istvan 107-108
de Vaucouleurs, Gérard 23-25
Dierickx, Marion 63
Digitized Sky Survey 2 70
Doeleman, Shep 114-115

EAGLE supercomputer 71
Einstein, Albert 20, 42, 48
    general theory of relativity 20, 42, 48
    theory of gravity 21
Energy 30, 65, 75, 140-141, 166
European Southern Observatory (ESO), Chile 11, 38, 45-46, 48, 52, 71, 89, 106, 108, 110, 119
    La Silla Observatory 46, 84, 89, 133
    Paranal Observatory 106, 110
European Space Agency (ESA) 51, 62-63, 100
    Darmstadt control room, Germany 102
    Gaia spacecraft 97, 99-105, 107, 150, 161-162
    Hipparcos satellite 100, 150
    Planck mission 29-30, 117-118

PLAnetray Transits and Oscillations (PLATO) satellite 145
Ewen, Harold 113
Extragalactic surveys 106-107

Finkbeiner, Douglas 52
Flagey, Nicolas 133
Friction 19, 160

Galactic archaeology 10
Galilei, Galileo 12, 98
Genzel, Reinhard 110
Gould, Benjamin 87
Gravity 10, 12, 18-21, 32, 70, 103, 136, 140-141, 155, 159, 164
Great Debate 1920 14-15

Harvard College Observatory 43
Harvard-Smithsonian Center for Astrophysics 39, 52, 63
Harvard University 113-114, 163
Hawking, Stephen 167
Heisenberg's Uncertainty Principle 28
Herman, Robert 29
Herschel, Caroline 13
Herschel, John 126
Herschel, William 13-14, 98, 142
Hertzsprung, Ejnar 90
Hertzsprung-Russell (HR) diagram of stars 90-91, 101-102, 149
Hewish, Antony 138
Hubble, Edwin 15, 21, 23-25, 60, 98, 158

Inertia 19
International Centre for Radio Astronomy Research, Australia 71
Interstellar Boundary Explorer (IBEX) 83
IRAM NOEMA Observatory 114

Kant, Immanuel 12, 14
Keenan, Olivia 66
Kepler, Johannes 16-19, 22, 141
    Laws of planetary motion 16-18, 34, 36, 154
Koposov, Sergey 103
Kormendy, John 38
Kourou spaceport, French Guiana 99

Laboratoire d'Astrophysique de Marseille 42
Large Hadron Collider 140
Laser Guide Stars (LGSs) 108, 110
Laser Interferometer Space Antenna (LISA) 35, 54
Laser Interferometry Gravitational- wave Observatory (LIGO) 54-55
    Hanford, Washington 54
    Livingston, Louisiana 54-55
Leiden University, Netherlands 92
Lin, Chia-Chiao (C.C.) 34
Lindblad, Bertil 36-37
    Lindblad resonances 36
Loeb, Avi 63-64 163-164
Lowell Observatory 158

Mackey, Dougal 70
Magellan, Ferdinand 67
Massachusetts Institute of Technology 34, 52
Mass 19

Mather, John 29
Max Planck Institute for Extraterrestrial Physics, Germany 48, 62, 110, 153
Mayans 97
Measuring distances in space 43, 100, 107, 125, 149-150
    dust extinction 125
McDonald Observatory, Texas 62
    VIRUS-W spectrograph 62
Melott, Adrian 81
Milligan, David 102
Mount Wilson Observatory, USA 14-15, 60
Müller-Sánchez, Francisco 166

NASA 51-52, 63, 116
    Chandra X-ray Observatory 50, 52, 115, 137, 140-141, 166
    Cosmic Background Explorer (COBE) 29
    Galaxy Evolution Explorer (GALEX) 63, 140
    Goddard Space Flight Center 120-121
    Transiting Exoplanet Survey Satellite (TESS) 145
    Wide-field Infrared Explorer (WISE) 57, 128
    Wilkinson Microwave Anisotropy Probe (WMAP) 30
National Radio Astronomy Observatory (NRAO) 45
Newton, Isaac 18-19
    calculus mathematical language 18
    laws of motion 18-19
    theory of gravity 18-19, 21
Nobel Prize in Physics, 2006 29
Norris, Henry 90
Noyola, Eva 153-154

Parsberg, Manderup 17
Parsons William (3rd Earl of Rosse) 14, 139
Penziass, Arno 29
Pontificia Universidad Católica de Chile 107
Popescu, Cristina 118
Prague Astronomical Clock 17
Ptolemy 16
Purcell, Edward 113

Radio astronomy 24-25, 113, 138
RECONS 92
Resistance 19
Rich, Michael 36

Sagan, Carl 80
Saglia, Roberto Philip 62
Shanghai Observatory 37
Shapley, Harlow 14-15
Shen, Juntai 37, 39
Slatyer, Tracy 52
Slipher, Vesto 158, 161
Smithsonian Museum of Natural History 14
Smoot, George 29
Space Shuttle 99
    Discovery 98
    Endeavour 98
Spectroscopic observations 46
Spizer, Lyman 115
Su, Meng 52
Swan Leavitt, Henrietta 43

Telescopes 14, 22, 28, 98, 119
    Allen Telescope Array, USA 113
    ALMA 112, 114
    APEX 114
    Aperture Spherical Telescope (FAST), Chia 113
    Arecibo, Puerto Rico 67, 113
    Atacama Large Millimeter/submillimeter Array (AMA) 111, 114, 127
    Auxiliary 11
    Blanco, Chile 36
    Effelsberg radio telescope, Germany 112, 150
    Event Horizon Telescope (EHT) 47-48, 114-115
    Extremely Large Telescope (ELT) 111, 119
    Fermi Gamma-ray Space Telescope 52, 116-117
    first 12
    gamma-ray 116-117
    Gran Telescopo Canarias, Canary Islands 119
    Greenland 114
    ground-based 24, 104, 110-111
    Herschel Space Observatory 62-63
    High Sensitivity Array, USA 150
    Hooker 15, 21, 60
    Hubble Space Telescope 32, 38, 42, 51, 60-61, 75, 84, 98-100, 110, 121, 127, 131-132, 134, 140-142, 147, 149, 153, 155, 161, 167
        Advanced Camera for Surveys (ACS) 99-100
        Cosmic Origins Spectograph (COS) 99
        COSTAR 98-99
        High Speed Photometer 98
        NICOMS 99
        Space Telescope Imaging Spectograph (STIS) 99
        Ultra Deep Field Image 12
        Wide Field Camera 3 (WFC3) 99, 100
    interferometry 113
    IRAM 30-meter, Spain 114
    James Clerk Maxwell Telescope 114
    James Webb Space Telescope (JWST) 120-121
    Keck Telescope, Hawaii 119
    Kepler Space Telescope 146
    Kitt Peak 114
    Large Area Telescope (LAT) 116
    Large Synoptic Survey Telescope (LSST), Chile 106, 118
    Leviathan 139
    Lovell Telescope, Jodrell Bank 113
    radio 66-67, 104, 112-113, 150
    reflective/mirrors 13, 18, 98, 106-107, 110-111, 119, 121
    Robert C. Byrd Green Bank Telescope 112
    Schmidt 71
    Sloan Digital Sky Survey (SDSS), New Mexico 42,103, 118-119
    South Pole Telescope, Antarctica 114
    space-based 24, 32, 62, 98-99, 102, 115, 140
    Spitzer Space Telescope 63, 103, 115, 116, 141